Defence Services Works Library

DWS

Please return books on or before the last date stamped below

---

19 MAY 1993

-3 ~~~ 1994

P.O. Box 1734, Rectory Road, Sutton Coldfield, West Midlands B75 7QB
Telephone: 021 311 ___2176___ Switchboard 021-378-1282

TRANSPORT RESEARCH LABORATORY
Department of Transport

STATE-OF-THE-ART REVIEW 4

ROAD AGGREGATES AND SKIDDING

by Roger Hosking

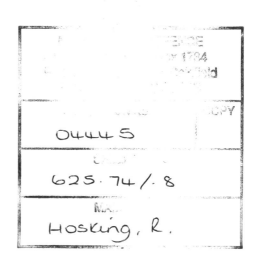

LONDON: HMSO

© Crown Copyright 1992
Applications for reproduction should be made to HMSO

First Published 1992
ISBN 011 551115 6

The views expressed in this review are not necessarily those of the Department of Transport

HMSO publications are available from:

HMSO Publications Centre
(Mail, fax and telephone orders only)
PO Box 276, London, SW8 5DT
Telephone orders 071-873 9090
General enquiries 071-873 0011
(queuing system in operation for both numbers)
Fax orders 071-873 8200

HMSO Bookshops
49 High Holborn, London, WC1V 6HB
(counter service only)
071-873 0011 Fax 071-873 8200
258 Broad Street, Birmingham, B1 2HE
021-643 3740 Fax 021-643 6510
Southey House, 33 Wine Street, Bristol, BS1 2BQ
0272 264306 Fax 0272 294515
9-21 Princess Street, Manchester, M60 8AS
061-834 7201 Fax 061-833 0634
16 Arthur Street, Belfast, BT1 4GD
0232 238451 Fax 0232 235401
71 Lothian Road, Edinburgh, EH3 9AZ
031-228 4181 Fax 031-229 2734

HMSO's Accredited Agents
(see Yellow Pages)

and through good booksellers

Contents

	Introduction	1
1	Studies of aggregates	3
	1.1 Early roads	3
	1.2 Research before the 1914-1918 war	3
	1.3 Creation of the Road Board	4
	1.4 The first British Standard Specifications	5
	1.5 The Ministry of Transport	5
	1.6 Research on aggregates for concrete	6
	1.7 Dr. Bernard Knight	6
	1.8 The formation of the Road Research Laboratory	6
	1.9 Early work of the Road Research Laboratory	7
	1.9.1 First projects	7
	1.9.2 Variability of aggregates	8
	1.9.3 "State of the art" in the late 1940s	9
	1.10 Post-war projects	10
	1.10.1 Work of the (Transport and) Road Research Laboratory	10
	1.10.2 Other research	13
2	Sources and production of road aggregate	14
	2.1 Sources of road aggregate in Great Britain	14
	2.1.1 Metallurgical slags	14
	2.1.2 Sands and gravels	17
	2.1.3 Igneous rocks	19
	2.1.4 Metamorphic rocks	20
	2.1.5 Sedimentary rocks	21
	2.1.6 Polish-resistant aggregates	23
	2.2 Production	28
	2.2.1 Crushing	29
	2.2.2 Screening	31
	2.2.3 Distribution	32
	2.3 Waste and low-grade aggregates	34
	2.3.1 General	34
	2.3.2 Colliery shale	35
	2.3.3 Pulverised fuel ash	36
	2.3.4 Incinerated domestic refuse	37
	2.3.5 China clay sand	38
	2.3.6 Spent oil shale	38
	2.3.7 Miscellaneous wastes	39

3	General requirements of roadmaking aggregates		41
	3.1	Practice in Great Britain	41
		3.1.1 Water-bound and unbound materials	41
		3.1.2 Concrete road construction	45
		3.1.3 Bituminous road construction	56
		3.1.4 Resin-bound surfacings	65
		3.1.5 Setts and pavers	65
	3.2	International recommendations for aggregates testing	67
4	Tests for roadstones and aggregates		73
	4.1	Test procedures	73
		4.1.1 General	73
		4.1.2 Sampling	76
		4.1.3 Tests for strength	80
		4.1.4 Density and water absorption	87
		4.1.5 Shape and texture	89
		4.1.6 Polishing resistance	93
		4.1.7 Compactability tests	94
		4.1.8 Sieve analysis	95
		4.1.9 Tests for soundness	98
		4.1.10 Chemical tests	100
		4.1.11 Other tests	101
		4.1.12 Classification	103
	4.2	Correlation between tests	107
		4.2.1 Petrology and other characteristics	107
		4.2.2 Correlation between mechanical tests	113
5	Special requirements of aggregates for skid-resistance		125
	5.1	Bituminous surfacings	125
		5.1.1 Development of the polished-stone value test	125
		5.1.2 Aggregates and polishing	129
		5.1.3 The effect on skid-resistance of factors other than PSV	130
		5.1.4 Mixtures of aggregates	131
		5.1.5 Relationship between polished-stone value and skidding	131
		5.1.6 Variation in skid-resistance at a road site	137
		5.1.7 Turning and braking	137
	5.2	Aggregates in concrete	137
		5.2.1 Accelerated wear machine	137
		5.2.2 The polished-mortar value test	140
		5.2.3 Black deposits	142
	5.3	Aggregates in other road materials	142
		5.3.1 Resin surfacings	142
		5.3.2 Other special polish resistant surfacings	145
	5.4	Mechanism of polishing of aggregates	147
	5.5	Road experiments	148
		5.5.1 General	148
		5.5.2 Results from small-scale road experiments	149
		5.5.3 Results from full-scale road experiments	149

6	Road surface characteristics		154
	6.1 Resistance to skidding		154
		6.1.1 General	154
		6.1.2 Road surface projection shape and friction	157
	6.2 Measurement of resistance to skidding		159
		6.2.1 General background to studies in Great Britain	159
		6.2.2 Small braking-force trailer	161
		6.2.3 SCRIM	162
		6.2.4 Portable skid-resistance tester	166
		6.2.5 The measurement of macro-texture	166
	6.3 Skidding accidents: causes and costs		169
		6.3.1 The causes of road accidents	169
		6.3.2 The magnitude and cost of road accidents	169
		6.3.3 Relation between skid-resistance and accident rate in Great Britain	170
		6.3.4 Some accident/skidding studies made overseas	171
		6.3.5 The saving of accidents and their cost by improvements in skid-resistance in Great Britain	174
	6.4 Traffic noise		174
	6.5 Reflecting properties of roads		175
		6.5.1 General	175
		6.5.2 Road surface reflection characteristics	175
7	Applications of research and experience		178
	7.1 Resistance to skidding		178
		7.1.1 Skid resistance standards for roads	178
		7.1.2 Polish-resistance standards for aggregates	185
	7.2 Overseas specifications for skid-resistance & aggregates		188
		7.2.1 Belgium	188
		7.2.2 Czechoslovakia	188
		7.2.3 Finland	188
		7.2.4 France	189
		7.2.5 Italy	189
		7.2.6 Japan	190
		7.2.7 The Netherlands	190
		7.2.8 Poland	190
		7.2.9 Spain	191
		7.2.10 Sweden	191
		7.2.11 Switzerland	191
	7.3 Specifications for aggregates		192
		7.3.1 National and International Standards	192
		7.3.2 A chronology of British Standards referring to aggregates	193
		7.3.3 A history of British Standard BS 812	194
	7.4 Current British Standards for aggregates and roadstone		198
	7.5 ASTM Standards for aggregates		199

8	Author's thoughts		200
	8.1 Sources and production of road aggregates		200
		8.1.1 General	200
		8.1.2 Super-quarries	201
		8.1.3 Mining	201
	8.2 General requirements of roadmaking aggregates		202
	8.3 Tests for roadstone and aggregates		202
		8.3.1 General	202
		8.3.2 Grading of aggregates	203
		8.3.3 Razor-blade shape	203
		8.3.4 Classification	204
	8.4 Special requirements of aggregates for skid-resistance		204
	8.5 Road surface characteristics		205
		8.5.1 The effect of tyre compound and tread pattern on the significance of skid-testing with SCRIM and HSTM	205
		8.5.2 Skidding on dry roads	208
		8.5.3 Further thoughts about texture requirements	209
		8.5.4 The effect of present-day trafficking on the PSV/SFC relationship	209
	8.6 Applications of research and experience		210
		8.6.1 Specification limits	210
		8.6.2 Skid-resistance maintenance policies	211
		8.6.3 Further saving of accidents and their cost by improvements in skid-resistance	213
		8.6.4 Tougher standards for tougher chippings	213
Acknowledgements			215
References			216
Appendix	Abbreviations and acronyms		228
Index			230

Introduction

This book aims to review the subject of road aggregates with particular emphasis on the characteristics of aggregates that relate to skid-resistance. Additional chapters are concerned with the measurement of skid-resistance and the various skidding "standards" that have been proposed.

The first Chapter "Studies of aggregates" outlines the development of our present knowledge of aggregates and roadstones from the early days of McAdam and Telford to the present time. An understanding of this history helps to give a clearer understanding of the present "state of the art" and could be of value in planning future studies. The depth of treatment of the various subjects has been tapered off as the Chapter progresses, because these later topics are dealt with in the main body of the book.

Chapter 2 gives an account of the sources of aggregates that are available in Great Britain and outlines the methods that are employed in their extraction and production. Particular emphasis is given to sources of aggregates that are of value in imparting skid resistance to roads. The final part of Chapter 2 is concerned with the utilisation of waste and low grade materials, a subject that is becoming increasingly important now that resources of higher quality materials are becoming depleted and there is more awareness of the need to conserve these resources.

The third Chapter is concerned with the general requirements of aggregates and roadstones. Some mention of requirements for roads in developing countries has been included because they could be of value in these areas as well as for roads used for forestry and agricultural purposes. The requirements laid down in a number of specifications such as the British Standards and the Department of Transport's Specifications are referred to in context. It should be borne in mind that these specifications undergo regular revision and that they are soon to be replaced by European Standards. Comprehensive treatment of "standards" for skid resistance and the requirements for aggregates needed to provide skid resistance is given in Chapter 7.

Chapter 3 begins with British practice and covers each of the main types of construction, including the more recent special resin-bound skid-resistant surfacing systems. This is followed by an account of the international recommendations that resulted from a review of aggregate testing methods used in many countries. However these recommendations can only serve as a general guide because the subject is still under active consideration.

The fourth Chapter of this book deals with test procedures. It begins with the problems of variability and reviews current attitudes to precision estimates. It continues with an account of what are considered to be the more important tests for aggregates. References are made to the various published specifications for these tests. This Chapter concludes with a collection of published correlations between the results of a number of different tests for aggregates.

The next two Chapters concentrate on skidding resistance and road safety. Chapter 5 deals with the part played by aggregates in imparting skid resistance to road surfacings. It concludes with a summary of some of the many road experiments that have been carried out with the object of studying the behaviour of aggregates under the abrasive and polishing action of traffic.

Chapter 6 is concerned with road surface characteristics. Resistance to skidding is the main theme, but reference is also made to the associated problems of traffic noise and light reflecting properties since these are also influenced by the choice of aggregate. A section of this Chapter gives an account of some of the work that has been done in other countries with regard to skidding problems. Methods of measuring skid resistance and surface texture depth are then described, particular attention being given to the terminology that is in current use. The final part of this Chapter deals with the causes of skidding accidents and quantifies the savings in accidents and their cost that can be expected by improvements in skid resistance.

The seventh Chapter deals first with the development of skidding standards in Great Britain and the associated standards for aggregates. This is followed by an account of the similar measures that have been introduced in other countries. Next comes a list of the various British Standards Specifications relating to the testing and use of aggregates, followed by a history of the development of BS 812 (the base British Standard for the testing of aggregates) to its present form.

Chapter 7 also includes reference to the present situation with respect to British and overseas standards for aggregates and concludes with lists of the current British and American Standards relating to aggregates.

Chapter 8 sets out some thoughts that have been generated whilst preparing this book. This is followed by a Bibliography of all the source publications for this book, except for the Standards listed in Chapter 7.

An Appendix lists the many abbreviations for organisations, tests and measurements that are commonly used in connection with aggregates and skidding.

The Geological Society's "Engineering Geology Special Publication No. 1" (Geological Society, 1985), now under review, is recommended for further reading on the subject of aggregates, particularly with regard to petrographical aspects, production and use outside highway engineering.

Another recommended publication is the Institution of Civil Engineer's guide "Wastes for imported fill" (Sherwood, 1987). This guide considers the extent to which waste materials and industrial by-products can be used as fill material in road construction, building work and land reclamation.

A valuable source of information on British, European and International standards is "Standards for aggregates" (Pike *et al*, 1990).

Studies of aggregates

1.1 Early roads

The main purposes of a road pavement are to protect the underlying layer from deformation under trafficking and to provide a comfortable and safe riding surface. Since large quantities of material are required for this purpose, it is usually essential that it should be of low cost. Both timber and stone have been used since very early times, either on their own or in combination with other materials, but stone has usually been preferred because of its greater durability. Stone has been used in two main ways, either as shaped blocks (setts or flags) or as an aggregate (crushed rock or gravel).

Although high quality highways had been built in Egypt and Babylonia centuries before, the first in Britain were constructed by the Romans. They used layers of aggregates bonded with mortar, the particle size diminishing towards the surface. Their wearing course consisted of dressed stone set in mortar on their more heavily trafficked roads and hoggin-like material (sand, gravel and clay mixtures) for the less important roads.

After centuries of neglect of the highways, growing prosperity in Britain led to the need for improved travel between our major cities, particularly during the winter months. The first step was taken in the 16th century when parishes were made responsible for the roads in their own localities. Next came the Turnpike Trusts that were created between about 1750 and the end of the 19th century to meet the growing need for better travel, and the resulting toll roads served the country for a while. Two important highway engineers emerged during this period, John McAdam (1756-1836) and Thomas Telford (1757-1834). Telford used a hand-pitched base of large stones surfaced with a regulating course of smaller stones. McAdam preferred to use compacted layers of smaller graded stones throughout his construction. In so doing he laid the foundations of modern flexible road design, and gave his name to a group of road materials.

By the end of the 19th century much of the traffic had diverted to the railways and canals and the Turnpike roads had deteriorated as a result of poor maintenance. This led to the Turnpike Acts being revoked in 1888 and the responsibility for the maintenance of major roads being transferred to the County authorities. These were required to appoint a Surveyor to supervise the road work; payment was through the county rates. It was during this period that the first recorded studies were made of the behaviour of roadstones under traffic.

1.2 Research before the 1914-1918 war

One of the major problems of early traffic was caused by the very high stresses that pavements had to withstand from rigid tyres. Even though there was legislation to limit the maximum wheel loadings, contact pressures could be as high as 0.5 tonne per 25mm width of tyre. Stone sett pavements were in general use in the cities. These were laid on a granular or cemented base and the stone surface was textured to give a good foothold to horses. The problems that arose led to studies being made by a number of engineers from the middle of the 19th century onwards. One of the pioneers was Walker who studied the relative wear of different kinds of rock in a machine of the grinding wheel type and related the results to their performance in full-scale road experiments under heavy traffic conditions in London. Another pioneer was Colonel Heywood whose road trials demonstrated that some of the hardest types of rock became polished after less than two years of trafficking. He then carried out further trials and developed a specification that gave a reasonable compromise between durability and skid-resistance.

These early road trials were followed by the work of E J Lovegrove, Borough Engineer of Hornsey, who devised an attrition testing machine. This was the first mechanical test to be used to assess a wide range of roadstones in Great Britain. An outline of the Lovegrove attrition test is given in Chapter 4. Starting in 1895, Lovegrove carried out a series of tests on a wide range of roadstones and related the results to the petrological characteristics of the rock types involved. He was assisted in this work by Dr John S Flett (later Sir John) of H M Geological Survey and J Allen Howe of the Museum of Practical Geology. Their work was first published in a series of articles in the "Surveyor" in 1905. This was followed by one of the few books that have been devoted entirely to the subject of roadstones, "Road-making stones. Attrition Tests in the Light of Petrology" first published in 1906. (Lovegrove *et al*, 1929).

In 1908 Mr Logan Waller Page, Director of the United States Department of Agriculture, became interested in Lovegrove's work. This led to a range of Lovegrove's samples of roadstones being subjected to a series of tests that were in use in the United States of America. These were the Dorry abrasion test - for hardness, the Page impact test - for toughness, and a cementation test - for the binding power of the rock dust. Further information on these tests is given in Chapter 4.

At the time of the beginning of Lovegrove's work, roads were mainly water-bound. However tar treatment soon became commonplace followed by the widespread use of coated macadam and other bituminous materials. Iron-shod horses drawing vehicles with steel tyres were also being replaced by motor vehicles with pneumatic-tyres.

1.3   Creation of the Road Board

The coming of motor vehicles led to a further reform in the administration of roads, namely the creation of the Road Board in 1909. The Lovegrove attrition test was considered to be of such national importance that it should no longer continue under private control. This led to the formation of the Road Board Laboratory as a division of the Engineering Department of the National Physical Laboratory, Teddington. The attrition machine was duly installed there in 1912.

The Lovegrove attrition test combined attrition with slight impact. Despite their early enthusiasm, the workers at the National Physical Laboratory (NPL) came to prefer the separate assessment of these properties. Attrition being measured by the Deval attrition test (invented by M Deval, the French engineer), and impact strength by the Page impact test (invented by Logan Waller Page).

1.4 The first British Standard Specifications

Lovegrove appreciated the need for uniformity in quarry output for the various types of roadstone. This led to the matter being referred to the newly formed British Engineering Standards Committee of the recently (1901) founded British Standards Institution and the eventual publication of British Standard Specification BS 63 "Broken Stone and Chippings" in August 1913. This was the first of the many British Standard Specifications that were to be published concerning roadmaking aggregates. BS 63:1913 contained an Appendix which set out 12 different "trade names" for groups of road-making rocks. These were formulated by the Geological Survey and Museum to provide a relatively simple classification for road making purposes, as compared with the great variety of rocks that had been described and named by geologists.

A list of the various British Standard Specifications relating to road aggregates is given in Chapter 7, together with a history of British Standard BS 812.

1.5 The Ministry of Transport

The coming of the war in 1914 accelerated the adoption of motor traffic for both commercial and private purposes and so led to the need for a more comprehensive transport plan. The Road Board was consequently dissolved in 1919 and the Ministry of Transport created to take its place. The testing work on roadstones was continued at the NPL, but the official title of the laboratory was changed to "The Ministry of Transport (Roads Department) Research Laboratory".

By 1924 the Ministry of Transport Research Laboratory was carrying out the first commercial tests on roadstones in Britain. The principal tests were:

i Abrasion, by means of a four-cylinder Deval machine.
ii Impact, by means of a Page impact machine.
iii Hardness, by means of a hardness machine of the Dorry type.
iv Cementation, by means of a ball mill for grinding the material, a forming machine for making the briquettes and a large impact machine to test them.
v Crushing strength by means of a compression testing machine.
vi Absorption of water test.
vii Microscopical examination.

All these tests were in general use by the mid-1930s.

Concrete road construction was not seriously contemplated until the formation of the Ministry of Transport. By this time (1919) extensive concrete pavements were being laid in the United States of America and, as a consequence, the Ministry of Transport investigated the feasibility of concrete roads in Britain by constructing a number of new roads in this way. These included the Colnbrook By-Pass where, under the provisions of the Roads Improvement Act 1925, the Roads Department of the Ministry of Transport undertook to construct and equip an Experimental Station at Harmondsworth, Middlesex. When this small experimental station was built in 1931 to test the concrete cubes associated with this work, it was little imagined that it would eventually become the present Transport and Road Research Laboratory.

1.6 Research on aggregates for concrete

At about this time the London County Council (later to become the Greater London Council) made many contributions to British specifications for aggregates in their Code of Practice. In particular, they formulated comprehensive requirements for concreting aggregates, which were later to form the basis of many of our national specifications.

Concrete mix design was also being actively researched at this time. Publications included L N Edwards' surface area method, the fineness modulus method of D A Abrams, the mortar voids method of A N Talbot, the "yield"/solid ingredients method of I J Mair and H N Walsh's grading method. (Edwards, 1918; Abrams, 1919; Mair, 1934; Walsh, 1933).

1.7 Dr. Bernard Knight

Dr. Bernard H Knight was another pioneer in the study of roadstone and aggregates. His "Road Aggregates. Their Uses and Testing" was published as Volume III in the "Roadmaker's Library" (Knight, 1935).

Knight was an enthusiastic and able petrographer who, unlike some of his contemporaries, firmly believed that it was the mineralogical rather than the chemical constitution of a rock that was significant in governing its behaviour in the road. He was also critical of the "trade group" classification of aggregates which had appeared in BS 63:1913, and proposed what he considered to be a more scientific classification. However his classification was never accepted by the British Standards Institution, which preferred to keep a rather simplified version of the BS 63:1913 groups.

1.8 The formation of the Road Research Laboratory

The work of this Ministry of Transport Station widened and in 1933 the duty of directing and supervising this work was transferred to the Department of Scientific and Industrial Research and the name changed to the Road Research Laboratory (RRL). The research work on roads which had been carried out at the NPL and also at the Ministry of Transport's Chemical Research Laboratory was then transferred to the RRL at Harmondsworth. This Laboratory was later to be transferred to Crowthorne and eventually grow to become the present-day Transport and Road Research Laboratory (TRRL).

For the discharge of their duty the Department of Scientific and Industrial Research appointed a Road Research Board to advise generally on the conduct of research into such matters as road materials and methods of construction. The Department of Scientific and Industrial Research became responsible for the issue of an Annual Report and scientific publications, but the Ministry of Transport remained responsible for full-scale tests under service conditions. These Annual Reports were issued from 1935 until 1939 and their issue was resumed again after the war in 1946. These publications contain a wealth of information and have made a valuable contribution to our knowledge of roads.

1.9   Early work of the Road Research Laboratory

1.9.1   First projects

Upon its formation, the RRL inherited the roadstone test facilities of the Ministry of Transport's Research Laboratory at Teddington. An early action was to make an appraisal of the research needs for roadstones and aggregates. This appraisal included the experience of surveyors and road engineers, the British Standard Specifications that were beginning to appear, the requirements of the Ministry of Transport, and a general consideration of the problems in highway engineering consequent to the great increase in motor traffic.

The first Annual Report of the Director of Road Research (Director of Road Research, 1935) made reference to the research needs on aggregates for road foundations, concrete and bituminous materials. Points specifically mentioned were:

i   The intrinsic properties of strength, deformability and attrition.
ii Grading.
iii The handling properties of plasticity and brittleness.
iv The needs of concrete, particularly grading.
v   The needs of bituminous materials, particularly grading, shape, adhesion of binder and surface area.

By this time there was general agreement among surveyors and road engineers that flaky and elongated material was undesirable in aggregates intended for road use. An example was the newly published BS 594:1935 "Rolled Asphalt. Fluxed Lake Asphalt" where it was required that the aggregate used "shall be angular but not flaky". One of the first projects of research was the development of tests for the measurement of elongation and flakiness (Markwick, 1936). These tests formed the basis of the British Standard tests for these properties, first published in BS 812:1938. Although more precise methods have been developed from time to time, they have been rather tedious to carry out and have not gained general acceptance in Great Britain as practical methods to be used under normal commercial conditions. Markwick's flakiness and elongation index tests survive virtually unchanged in BS 812. The elongation index test has fallen out of general use but is still used for special purposes. (See Chapter 3.1.2.5)

By 1937, the effect of crushing methods on the shape of chippings was being studied and also the interlocking properties of aggregates. Another project was the investigation of the grading requirements of single-sized chippings. In this connection, contemporary published records of twenty gradings of stone that had been used successfully were examined, and it was found that only one complied with the existing British Standard Specification! At that time the British Standard for single-sized chippings (BS 63) called for a content of not less than 70 per cent stone of the specified size and no oversize was permitted. The RRL studied the problem and put forward proposals that embodied the following principles:

i   A small percentage of oversize should be permitted.
ii   The percentage of the specified size required should be diminished with the size of the stone.
iii   The grading of the undersize material should be specified in such a way as to ensure that the majority should be in the next largest size to that specified.
iv   The amount of dust should be controlled.

These principles were adopted by the British Standards Institution and have been embodied in later editions of BS 63. (See also Chapter 3.1.3.)

Other research by the RRL at this time was concerned with sieving, sampling methods, the shape of gravels, variation within bulk material, surface area, and the effect of shape and size on the rate of spread of aggregates used in surface dressings. Studies were also made of a number of tests for mechanical strength. These included a modification of the Deval attrition test, a cylinder crushing test (later named the aggregate crushing test) and the Los Angeles abrasion test. The relevant findings are included in later Chapters.

1.9.2 Variability of aggregates

Much progress had been made by 1939. Considerable variation had been found in the grading of aggregates in railway trucks. Much of the apparent variability shown by test results was found to arise from limitations in sampling and sieving which caused even a uniform material to appear variable. These differences were found to be produced by variations in the size of test sieve apertures, by subjective variations in sieving, and by errors due to non-representative sampling. Small differences in sieve aperture within the tolerances permitted by the British Standard BS 410:1931 were shown to have a surprisingly large effect on the grading. No advantage was found to be gained by the use of large numbers of sieves to specify the grading of aggregates, when the permitted differences in sieve apertures and the other causes of variation allowed large apparent differences in grading. Consequently it was recommended to the British Standards Institution that only 20 sieves should be used for gradings covering the whole range from 200 mesh (75 µm) to 3 inch (75 mm), and that for many purposes the complete range could be covered by 11 sieves only.

Special work was mostly undertaken by the RRL during the 1939-1945 war. One activity was to develop the aggregate crushing test to the stage where it could provide a quick assessment of aggregates for constructional work. This led to the adoption of this test by the British Standards Institution, rather than its rival, the Los Angeles abrasion test. (See Chapter 4.1.3.6.)

1.9.3 "State of the art" in the late 1940s

A landmark in the history of aggregates was the publication of Road Research Special Report No. 3 (Phemister *et al*, 1946). Apart from setting out the "state of the art" at the time it drew attention to the need for many improvements in knowledge and specification. These included:

i   The importance of repeatability and reproducibility. Figures quoted showed that the results of the tests on aggregates were much more reproducible than tests made on single pieces of rock. The precision of one aggregate crushing or Los Angeles test was found to be comparable to that of the mean of 90 Page impact tests or 60 crushing strength tests.

ii The need for studies of the correlation between the various tests. This led to the discovery of almost numerical equality between the aggregate crushing and the Los Angeles tests; only poorly defined relationships were found between the other tests. (See Chapter 4.2).

iii Simplification of the trade groups of BS 812 and improvements in the way BS 812 specified petrological name and description (hardness, colour, grain, imperfections, etc.). (See Chapter 4.1.2.)

Special Report No.3 also differentiated between two kinds of mechanical tests on roadstone. One kind consisted of the tests of the quality of the rock, which were made on specially prepared test specimens and in general could not be applied to the usual sizes of commercial aggregates. The other consisted of the tests which were applied to the aggregate as supplied by the quarry or pit and which, therefore, in addition to giving a measure of the quality of the rock, must also necessarily be influenced by the size and shape of the particles.

The tests on prepared specimens in vogue at the time were the crushing strength test, the Dorry abrasion test and the Page impact test, all of which were specified in BS 812:1943.

Other commonly used tests on aggregates were the Deval attrition test (both wet and dry versions), the Los Angeles abrasion test, the aggregate crushing test, and tests for specific gravity, water absorption and cementation value. These tests appeared in BS 812:1943 except for the Los Angeles abrasion test, which was carried out in accordance with American Standard ASTM C131-39 and has never been included in a British Standard Specification, and the cementation test which, although in considerable use at one time, was finding little application by 1946. All these tests are described in more detail in Chapter 4.

Technical Paper No. 10 (Shergold, 1948) supplemented Special Report No. 3, giving a review of available information on the significance of roadstone tests. It included much information on the correlation between the various tests (see Chapter 4).

The main conclusions drawn by Shergold were:

i   Some correlation existed between the results of attrition, abrasion, impact and crushing strength tests on roadstone, although the results on one test could not be predicted from those of another except within very wide limits.

ii   The service behaviour of aggregates could be related to their petrological characteristics, but specialised knowledge was required for an understanding of this subject.

iii   Tests on prepared individual specimens of rock in general showed poor correlation with the behaviour in service of the aggregate prepared from the rock.

iv   Tests on the aggregates rather than on individual specimens in general showed good correlation with their service behaviour, but existing knowledge justified only a tentative application of specification limits.

v   The British Standard aggregate crushing test gave a satisfactory measure of the crushing resistance of an aggregate.

vi   It was desirable that British Standard tests be developed to measure the resistance of aggregates to abrasion and impact.

vii   The tests on individual specimens should be retained until a wider experience of tests on aggregates was available.

viii In order to extend existing knowledge of the significance of tests results, a large number of aggregates should be tested at the time of laying and the service records of the roads collected over a period of years.

These conclusions led to the development by the RRL of our present-day aggregate impact test and aggregate abrasion test, and to the study of the behaviour of aggregates in a large number of full-scale and small-scale road experiments.

1.10 Post-war projects

1.10.1 Work of the (Transport and) Road Research Laboratory

Because most of the post-war work of the (T)RRL is covered in detail in later Chapters of this book, only a brief mention of the more important subjects is made in this Chapter. However more detail is given for a few subjects that are not so covered, but are of historical interest.

The first list of sources of supply of road aggregates was published in 1948 (Road Research Laboratory and Geological Survey and Museum, 1948). It included the name of owner or occupier, the name and situation of the quarry, the rock type, the trade group and the colour of the rock. It listed both quarries and pits and was sub-divided into counties. Revisions were published at intervals by the RRL and Geological Survey and Museum for many years: this function is now performed by the British Geological Survey's "Directory of Mines and Quarries" (British Geological Survey, 1985).

The first of the RRL's full-scale road experiments was laid shortly after the war to study the behaviour of road materials under service conditions. Some were to examine the behaviour of local aggregates such as crushed quartzite gravels in surface dressings at Rowsley in Derbyshire and in experimental "thin surfacings" on the Stafford-Cannock road. Other road experiments particularly relevant to the problems of aggregates and skidding are described in later Chapters.

A full-scale crushing and screening plant was installed in Somerset County Council's Underwood quarry, in order to study the characteristics of crushing and screening plant.

Tests on twelve finishing crushers were completed by 1956. A summary of the findings of more general interest is given in Chapter 2. The detailed information obtained was published in Technical Paper No. 44 (Shergold, 1959), and the methods developed for assessing the performance of rock granulators under controlled conditions were published in Road Note No. 37 (Road Research Laboratory and Military Engineering Experimental Establishment, 1965).

A machine was constructed in the 1950s to allow the separate study of the different factors that cause wear on a surface dressing. Development of this accelerated wear machine led to our present-day polished-stone value test. (See Chapter 4.1.6.)

A landmark in the knowledge about roadmaking sands and gravels was the publication of Technical Paper No. 30 (Shergold, 1954b). This gave the results of a study of single-sized aggregates for roadmaking made in collaboration with the Ballast, Sand and Allied Trades Association. (See Chapter 2.1.2.3.) This work led to the publication of BS 1984:1953, "Single-sized gravel aggregates for roads".

Research into methods of measurement of the efficiency of screens was commenced in 1954, and studies began of a single deck vibratory screen installed at the Laboratory. This work included the development of formulas for screening efficiency. Investigations were extended to working quarries and methods were developed for measuring the efficiency and other characteristics of screening plant under normal operating conditions, without interfering with production. Further information is given in Chapter 2.

At the request of the British Limestone Federation, tests were carried out to establish whether good-quality limestone aggregates can produce concretes of higher crushing and flexural strengths than those obtained with good-quality igneous rocks. (See Chapter 2.1.5.1.) An investigation was also made, with the assistance of the British Limestone Federation, into the road and traffic conditions under which different types of limestone could be used in the wearing courses of roads without polishing sufficiently to cause a slippery road surface. (See Chapter 4.2.1.6.)

An investigation was made of the extent to which the resistance to crushing, abrasion and polishing of different samples of roadstone from a number of sources varied. This led to a further study of sampling procedures involving a large sampling programme from the rock face and from the finished aggregate at seven quarries. A range of processing plants and types of rock were included. The main findings led to improvements in the British Standard methods of sampling aggregates. (See Chapter 4.1.2.)

A major project that began in the mid-1960s was concerned with blastfurnace slag of low density (Director of Road Research, 1965; Hosking, 1967a). In the previous decade a number of large blastfurnaces had come into operation which used high-grade imported iron ores. The slag produced from these furnaces usually failed to comply with the requirements in force at the time. (See Chapter 2.1.1.4.)

Preliminary work by the Tropical Section of the RRL led to a more detailed general study of the particular problems of unsound and low-grade aggregates. These problems are particularly acute in some tropical countries, but they are by no means limited to them.

In order to establish values to be included in the Ministry of Transport's Specification for Road and Bridge Works (Ministry of Transport, 1963) an investigation was made of the resistance to skidding at many of the RRL's full-scale and small-scale road experiments and at skidding accident sites. At the same time a survey was made of sources of stone of higher polish-resistance, which were required for the more exacting sites. This indicated a shortage of the best stones and necessitated a relaxation in the recommended values from those obtained from the investigation. These values were issued in Technical Memorandum T2/67 (Ministry of Transport, 1967). (See Chapter 7.1.2.)

High-quality calcined bauxite from Guyana, as used in the refractory industry, was found to maintain an exceptionally high resistance to skidding when used as a 3 mm grit in resin based binders even under the most severe traffic conditions (James, 1960; James, 1963). (See Chapter 3.1.4.)

The shortage of aggregates of high resistance to polishing led to intensive research into both finding new sources of naturally occurring and artificial polish-resistant materials. (See Chapter 2.1.6.)

Multiple regression analysis of available data allowed a formula to be developed relating the skid resistance of bituminous surfacings to the polished-stone value and traffic density (Szatkowski and Hosking, 1972). Details of this and the way in which it was used to implement skidding standards is dealt with in Chapters 5 and 7 respectively.

Another major project of the RRL in the late 1960s was stimulated by the disaster at Aberfan, public attention was focused on methods of removing the waste heaps that exist close to the mine workings. An obvious outlet was to use these materials in road construction. Studies showed that the material could be used for this purpose. The results of this and studies of other waste materials are described in more detail in Chapter 2.3.

A major project that started in the late 1960s was the collaborative research of the Sand and Gravel Association (SAGA) and (T)RRL. This included a study of unbound materials made with gravel aggregates. It led to a detailed investigation into the the use of the shear-box test for measuring the resistance to shear of a compacted mass of clean, graded aggregate as used in Type 1 sub-bases. A detailed investigation was also carried out into the compaction of aggregates using the British Standard vibrating hammer test (BS 1377). (See Chapters 3.1.1.3 and 4.1.7.)

Other collaborative work by the TRRL and SAGA included an investigation of the drying of flint gravels in commercial plants and a survey that showed that limestone-gravel bituminous materials do not require the addition of Portland cement as a precaution against stripping. (See Chapter 2.1.2.4.)

The resistance to skidding of concrete was the subject of considerable joint research by the TRRL and the Cement and Concrete Association. (Chapter 5.2.)

Collaborative work was also carried out by the TRRL with Asphalt and Coated Macadam Association (ACMA). (Chapter 3.1.3.)

Studies of the distribution of aggregates are outlined in Chapter 2.2.3.3.

1.10.2 Other research

Numerous contributions to our knowledge of aggregates and their use have been made both in Great Britain and overseas, by a large number of organisations and individuals. Reference to some of the overseas work is made in Chapter 3.

British contributions come from our Universities, the Institute of Geological Sciences, Local Authorities, the Geological Society, the Aggregates Producers and their Associations, the producers of road materials and many more.

Reference is made to as much of this work as possible in the following Chapters, but limitations of length make it impossible to give full coverage.

Much of the recent progress has been made possible by the general co-operation of the many representatives of interested organisations on the various British Standards Technical Committees dealing with aggregates and their use. Of particular importance have been the following:

i   Precision studies have been made of as many tests as possible. These have led to the improvement of some test methods and the revision of the procedure for others.

ii   There has been adjustment of many requirements to allow the greater use of local or alternative aggregates for many purposes. Conversely the requirements for skid-resistance have been tightened because increases in traffic have made it necessary.

2 Sources and production of road aggregate

2.1 Sources of road aggregate in Great Britain

In Great Britain, solid rock suitable for the manufacture of road aggregates is almost exclusively quarried from formations of the Palaeozoic and Pre-Palaeozoic geological ages. Few of the later formations are sufficiently indurated (strengthened by processes such as heat and pressure) to be of value for this purpose. This means that there are few quarries in the south east of the country. However this deficiency has been to some extent made good by the occurrence of flint gravels in this area. Gravels are also abundant in other parts of the country.

Other major sources of aggregates are the by-products of metallurgical and other processes.

A detailed account of all the various sources of aggregates will not be given, only a summary of the main features of each type and an account of some of the special investigations that have been made that relate to specific aggregate types.

Information on sources of supply of road aggregates is included in the British Geological Survey's "Directory of Mines and Quarries" (British Geological Society, 1985). This supersedes an earlier publication by the Road Research Laboratory and Geological Survey and Museum (Road Research Laboratory and Geological Survey and Museum, 1948-1968)

2.1.1 Metallurgical slags

2.1.1.1 General

Slag is a by-product (sometimes a waste-product) of a range of metallurgical processes, which can vary with both the material processed and the process itself. Because of this, its suitability for purposes such as the production of aggregates can vary from time to time. "Blastfurnace and Steel Slag: Production, Properties and Uses" (Lee, 1974) gives an excellent review of the uses of slags up to that time. However, changes in the manufacture of steel have already had a significant effect on the availability and nature of slags that can be used to manufacture aggregates. These, and the possibility of further changes, make any "state of the art" presentation involving these types of aggregates likely to be subject to change.

2.1.1.2 Types of metallurgical slag

Metallurgical slags have been used as a source of road aggregate from early times, for example, iron-smelting slag aggregate has been found at Worcester dating from Roman times. However, it was not until the coming of blastfurnaces in about 1830 that large quantities of slag were produced, but little was used in road construction until the 20th century. This led to the accumulation of large waste tips. Blastfurnace slag was at first the by-product of iron production from local low grade iron ores and the quantity produced was large in relation to the quantity of iron produced (more than one tonne per tonne of iron).

This earlier blastfurnace slag was of high density and became a premium source of road aggregate; a range of standards and specifications were drawn up for its use. However in the mid-1950s there was a change in production methods using the more economic higher grade imported iron ores. The consequences were an increase in coastal iron works, the closure of many of the inland works, and the production of slags of lower density. The quantity produced was also much reduced relative to the amount of iron (only about 0.25 tonnes per tonne of iron). This led to a change in the blastfurnace slag aggregates being marketed, some being entirely the newly-produced lighter-weight slag and others a blend of old slag from the tips combined with the new slag. More recently the methods of steel production have changed yet again and much steel is produced directly from iron ore without the intermediate production of pig iron.

Steel slag had been produced only in relatively small quantities until this recent change. Nevertheless it has been an important local source of aggregate and sometimes achieved premium status because of good polish-resistant properties.

Other slags have also been used for the manufacture of roadmaking aggregates, but they have only been available in relatively small quantities. These include phosphorus slag and, to a lesser extent, copper slag and zinc slag.

Blastfurnace slag has also been processed to manufacture a number of materials such as granulated slag for cement manufacture and slag wool for insulation. Another aggregate so produced is foamed slag for use as a light-weight aggregate. Cast slag products have also been used in road construction.

2.1.1.3 Dense blastfurnace slag

The earliest standards and specifications for slags were based on experience with dense blastfurnace slags. Large quantities of first class road aggregate were being produced; in many ways it resembled natural rock but a number of different problems were sometimes encountered. These were mainly due to the instability of the constituents of some slags. This gave rise to a separate set of standards and specifications from those for natural rock for both concrete and bituminous road materials. Knight included a chapter in his book on the properties and testing of slag and clinker (Knight, 1935). He appreciated that the chemical composition was important and refers to three tests that were being applied at the time for:

i   Lime/silica ratio - which should be less than 1.3.
ii   Sulphur content - which should not exceed 2 per cent.
iii   Weathering.

He had little faith in the rapid weathering tests of the time and recommended the following:

i   The slag was to be iron blastfurnace slag only.
ii   The slag was to be obtained from heaps which have been exposed to the weather for at least ten years.

iii The slag was to be crushed, screened and left to the action of the weather for at least twelve months before being coated with binder.

Tests for properties specific to blastfurnace slag were at one time confined to the British Standard specifications relating to the uses of such slag. BS 1047 for concreting aggregates was such a specification and included tests for lime unsoundness, iron unsoundness, sulphur content and acid soluble sulphate content. These tests are now being incorporated into various Parts of BS 812. Limits for total sulphur and acid soluble sulphate were maxima of 2% and 0.7% respectively. Blastfurnace slag aggregates for bituminous materials did not include an acid soluble sulphate limit and the permitted amount of total sulphur was relaxed to 2.75% maximum.

2.1.1.4 Lighter-weight blastfurnace slag

The changes in blastfurnace practice in the mid-1950s led to the production of slag that failed to comply with the requirement of BS 1047 for a minimum bulk density of 78 lb/cu.ft. (1250 kg/cu.m.), However there was evidence that it could be used successfully for road making. The problem was therefore investigated by the Road Research Laboratory in collaboration with the British Slag Federation (Director of Road Research, 1964).

A survey of examples of surfacings of various types where the lighter-weight slag had been used was made in London, the Southern Counties and South Wales. Some 80 sites were visited and the performance of the material in bituminous mixtures was assessed. It became clear from this survey that, provided an increased binder content (by weight) was used, the lighter-weight slags investigated could give a performance comparable with that from a conventional dense blastfurnace slag.

A further study (Hosking, 1967a) was made with the object of obtaining information on the properties of slag currently produced (1964-5). Samples were taken at hourly, daily and monthly intervals at each of ten works representing a cross-section of production at that time. These samples were tested for bulk density, particle density, water absorption, sulphur content, acid-soluble sulphate content, and resistance to abrasion, crushing and polishing. The results of the tests led to the following findings.

i   As expected, slags with low bulk densities tended to have lower resistance to abrasion and crushing, and higher resistance to polishing.
ii   There was a positive correlation between bulk density and particle density and the lighter slags tended to have a greater water absorption.
iii The bulk density increased as the particle size decreased, this being particularly noticeable with the lower density slags.
iv   The variability between hourly samples was relatively small; the daily samples showed greater variability, and the greatest variability occurred between the monthly samples.
v   The variability appeared to be independent of the type of blastfurnace slag.

2.1.1.5 Light-weight blastfurnace slag

Light-weight slag (foamed slag) is manufactured by treating molten slag with a controlled amount of water. This has the effect of expanding the slag into large masses having a cellular structure. Coarse aggregates so produced have a bulk density in the range 560 to 720 kg/cu.m.

(35 to 45 lb/cu.ft.). The corresponding figures for fine aggregate are 800 to 950 kg/cu.m. (50 to 60 lb/cu.ft.).

The value of light-weight slag lies in the production of light-weight concrete for building and bridge construction: it has no special value as a road pavement aggregate. BS 877 specifies requirements for foamed slag.

2.1.1.6 Steel slag

Until recently, steel slag has been produced in only relatively small quantities. It usually has a high lime content and therefore tends to contain unstable material unless it has been weathered for a considerable period of time (about a year). Steel slag tends to be denser and stronger than blastfurnace slag and some sources have provided an aggregate with a high resistance to polishing.

The present-day production of relatively large quantities of steel slag from the direct steel process has led to investigations of its properties. These have been mainly directed towards the problems of unsoundness due to the high lime content and the steps, such as weathering, that are needed to allow the production of satisfactory aggregates. These investigations have led to special requirements being formulated such as those for coated macadam (BS 4987:1988), where steel slag is required to have been weathered until it is no longer susceptible to "falling" (spontaneous degradation) and must have a compacted bulk density lying between 1,700 kg/cu.m. and 1,900 kg/cu.m. (106.5 lb/cu.ft. to 119 lb/cu.ft.)

2.1.1.7 Other slags

The only other slags of any importance are those of phosphorus, copper and zinc. In the mid-1960s production of phosphorus slag was as high as 100,000 tonnes per annum, but has subsequently declined. Phosphorus slag aggregate was of an attractive pale blue colour and had mechanical properties that made it suitable for road surfacing. Unfortunately it also tended to polish rather readily under traffic and therefore was only suitable for the surfacing of very lightly trafficked roads. Copper slag and zinc slag have also been used as a source of aggregate, but the small quantity has not justified a major study.

2.1.2 Sands and gravels

2.1.2.1 General

Sand and gravel beds tend to consist mainly of harder and sounder material. This is because they are derived from the weathering and transport of material from earlier formations. For this reason, too, they tend to consist of rounded particles rather than angular ones as is the case with crushed solid rock. However finished aggregates usually contain a proportion of angular particles as a result of the crushing of larger particles in the preparation plant. The proportion of crushed material in a gravel aggregate can significantly affect its properties. In particular an increase in crushed material will reduce workability and increase skid-resistance properties.

Britain is fortunate in that regions that are lacking in solid rock usually have abundant sand and gravel supplies. This is particularly the case in the south east of England where, in the London Basin and elsewhere, there are extensive deposits of flint that has been weathered from the Chalk. Extensive quartzite gravel deposits occur elsewhere; examples are the Triassic Bunter pebble beds of the north Midlands. Triassic limestone gravels are also important in some areas. In Scotland there are abundant glacial deposits consisting of material of various origins; examples are found in the Midland Valley.

In recent years there has been extensive exploitation of marine deposits of sands and gravels. These are dredged from many off-shore banks, particularly off the south-east coast. "Marine Dredging for Sand and Gravel" (Nunny and Chillingworth, 1986), "Sea Dredged Aggregates in Concrete" (Gutt and Collins, 1987) and "Marine Dredged Aggregates" (BACMI, 1987) are recommended for further reading.

2.1.2.2 Characteristics of sands and gravels

Historically, sands and gravels have been mainly used for making concrete whereas crushed solid rock aggregates have been preferred for bituminous mixtures. The reasons for the preferential use of sands and gravels in concrete are that their roundness is favourable for good workability and their natural grading is such that much "as-dug" material could often be used for making concrete without any further processing.

Providing their characteristics are understood, and a few precautions taken, there is no reason why gravels may not be used in bituminous mixtures. A difficulty that can arise is stripping of the binder with gravels such as flint. This problem can be overcome by adding hydrated lime or by the use of a suitable anti-stripping agent. Another characteristic of gravels is that the presence of a large proportion of rounded material tends to give a lower resistance to polishing; crushed material from the same source can often give a significantly higher resistance.

Early specifications for concreting sands tended to favour those with a relatively coarse grading. Whilst it is true that these are easier to use and do not require quite so much cement, those with a finer grading can also be used to make good quality concrete. The former rather rigid requirements of the British Standard (BS 882) have now been amended to allow the use of sands of a greater range of grading characteristics. This is particularly important in regions where finer sands predominate and where the sources of coarser sands are becoming exhausted.

The advent of marine dredged sands and gravels has heightened a problem that used to be rare. That is the deleterious effect of chloride ions on the steel reinforcement of concrete and nearby steel structures. Marine dredged material can contain excessive amounts of sodium and other soluble chlorides. A test for these now appears in the latest revision of BS 812.

Another problem that occasionally occurs with gravel aggregates is that of alkali-silica reaction, leading to premature failure of concrete. (See Chapter 3.1.2.3).

2.1.2.3 Studies of sands and gravels

A landmark in the knowledge about roadmaking sands and gravels was the publication of Technical Paper No. 30 (Shergold, 1954b). This gave the results of a study of single-sized aggregates for roadmaking made in collaboration with the Ballast, Sand and Allied Trades

Association. By this time 35 million tonnes of gravel and sand were being produced annually in Great Britain and changes in highway engineering practice had created a demand for "single-sizes" that could be used either singly or in combination to give consistent gradings for coated macadams, concrete, granular bases and rolled asphalt. The results of tests on 294 samples were related to the grading and shape requirements of the contemporary British Standards specifications for single-sized aggregates (BS 63:1951: BS 882:1944), and the strength requirements of BS 882:1944 (aggregate crushing value not more than 30) and BS 594:1950 (aggregate crushing value not more than 25).

This work led to the publication of BS 1984:1953, "Single-sized gravel aggregates for roads", which gave requirements for 1 1/2 inch, 1 inch, 3/4 inch, 1/2 inch, 3/8 inch and 1/4 inch single sizes (these approximate to the present-day single sizes of 37.5 mm, 28 mm, 20 mm, 14 mm, 10 mm and 6.3 mm respectively).

2.1.2.4 Bituminous materials made with limestone gravel aggregates

A survey of a number of roads carrying heavy traffic has been made (Road Research Laboratory, 1968) with the object of studying the condition and properties of bituminous materials made with limestone gravel. Particular reference was made to the resistance of these materials to the disintegrating action of water. The survey showed that limestone-gravel bituminous materials do not require the addition of Portland cement as a precaution against stripping in the presence of water.

2.1.2.5 The drying properties of flint gravels

An investigation was made by the Sand and Gravel Association's research team at the Transport and Road Research Laboratory, of the difficulty of drying flint gravels relative to other aggregates (Pike, 1971). The drying properties of gravels with different water absorption values were also investigated. It was concluded that, whereas flint gravels are more difficult to dry than some aggregates, they are not necessarily the worst materials in this respect; nor are weathered flint gravels, which tend to have a high water absorption, more difficult to dry than those exhibiting lower water absorption values. It was also concluded that improvements in drying techniques could be made, and that the introduction of a limit on residual moisture contents of aggregates used in dense bituminous mixtures should be considered.

2.1.3 Igneous rocks

Igneous rocks are those derived from molten material. The more important sources of aggregate in this category are members of the basalt, porphyry and granite groups, the finer-grained members tending to be the more useful as a source of road aggregates.

2.1.3.1 Basalts

The Whin Sill, consisting of Permo-Carboniferous intrusive quartz-dolerite sheets, provides an important source of aggregates in the north of England from north Yorkshire northwards. Basalts also occur in Derbyshire, Shropshire, near Birmingham and parts of Wales, and other basic igneous rocks in Cornwall, Devon, Somerset and Warwickshire. Basalts and dolerites provide important sources in many parts of Scotland.

This group of rocks is an important source of roadmaking aggregates. They are strong and many have a high resistance to polishing. However, a few sources, particularly in Scotland, contain olivine which has decomposed to clay. This can adversely affect mechanical strength but is more commonly manifest as high drying shrinkage characteristics that can lead to problems in concrete.

2.1.3.2 Granites

The granite formations in Leicestershire provide an important source of aggregates. Other important sources occur in the Malvern Hills, Eskdale, Cornwall, Devon, Anglesey and the Channel Islands.

Granites are usually strong; the finer grained deposits tend to have greater strength than the coarse grained granites that find favour for building and monumental work. Their resistance to polishing is usually good. Being acidic, they can call for greater attention to anti-stripping treatment than the more basic igneous rocks.

2.1.3.3 Porphyries

Formations of porphyry have provided some excellent aggregate sources in Wales and Scotland. Others occur in the North of England, Leicestershire, Somerset. Cornwall and Devon. Porphyries provide good all-round roadstones.

2.1.4 Metamorphic rocks

Metamorphic rocks are those that have been changed by the action of heat and/or pressure. Those that are of importance as a source of aggregate are hornfels, metamorphic quartzites and, to a lesser extent, schists.

Hornfels rocks have provided the strongest aggregates and were in great favour for road surfacing until the contribution of aggregates to skidding was identified. In quite recent times they have been in demand in some countries because they withstand the action of studded tyres better than any other aggregate. Hornfels has been quarried in Anglesey, Cornwall, Cumbria, Devon and, further north, in Dumfries & Galloway, Grampian and Shetland

Apart from the a poor resistance to polishing, hornfels rocks provide excellent road aggregates.

Quartzites have been important sources of aggregates in the Nuneaton area. Others occur in Shropshire. Apart from a tendency to poor adhesion to bituminous binders, quartzites provide a good road aggregate with a good resistance to polishing.

2.1.5 Sedimentary rocks

2.1.5.1 Limestones

a.   General

Strictly speaking, limestones are sedimentary rocks (laid down as a sediment) consisting largely of calcium carbonate. However the limestone group of road aggregates can also includes rocks with a substantial proportion of magnesium carbonate, examples are the Magnesian limestones. This group of rocks has provided an important source of aggregates in England and Wales and, to a lesser extent, Scotland. The extensive deposits of the Carboniferous age have been the major source, but Devonian and Silurian sources also exist. Some more recent deposits have also been exploited particularly in regions devoid of harder solid rock. The main occurrences of Carboniferous limestone are in the Mendip Hills, in Avon, north and south Wales, Derbyshire, north Yorkshire, Lancashire and Cumbria. Important sources of Magnesian limestone are found in Derbyshire, Nottinghamshire and Durham.

The more recent limestones tend to lack strength and abrasion resistance and are not resistant to damage by frost. A high susceptibility to polishing rules out most limestone aggregates from virtually all bituminous surfacings, but there are a few examples that have good resistance to polishing. These owe their polish-resistance to the presence of hard mineral particles. Some studies of particular interest are outlined below.

b.   Limestone aggregates in concrete

At the request of the British Limestone Federation tests were carried out by the Road Research Laboratory to establish whether good-quality limestone aggregates can produce concretes of higher crushing and flexural strengths than those obtained with good-quality igneous rocks (Director of Road Research, 1958). Six limestones from different parts of the country were compared with three igneous rocks of known good quality. All the limestones were used in mixes with a natural sand and, in addition, three were used with limestone fine aggregates.

It was deduced that for mixes of the same workability there was no advantage in using limestone fine aggregate. Small differences were observed between the strengths obtained with the various limestones, but there was no general tendency for concretes with the limestones as coarse aggregates to have different flexural strengths from those with the igneous rocks. Thus no advantages would be expected from using limestone aggregate in concrete-road construction other than those that might result from its lower thermal expansion compared with other types of aggregate.

c.   Polishing resistance of limestones

An investigation (Director of Road Research, 1960) was made with the assistance of the British Limestone Federation, into the traffic conditions under which different types of limestone could be used in the wearing courses without polishing . Samples of limestone from 19 sources were tested in the laboratory for resistance to polishing, and road surfacings with 11 of the limestones were tested to determine their resistance to skidding. After some apparent anomalies arising from the variability of some of the limestones had been investigated and satisfactorily explained, it was found that the behaviour of the limestones fitted into the general pattern of other types of stone. The laboratory test used in the assessment, a fore-runner of the polished-stone value test, gave as good an indication of the relative merits of limestones as of other stones.

2.1.5.2 Gritstones

a.   General

In strict petrographic terms, gritstones are coarse sandstones. However the gritstone group of roadmaking aggregates includes all sandstones (rocks consisting of cemented hard sand-sized particles) and similar rock types. In early times, when hand labour was used to manufacture roadstone, gritstones were a common source. They provided an easy source of high grade flags and setts for road construction because of the ease of working. However, with the advent of requirements for crushed stone and the use of machines (crushers, screens, etc.) to manufacture roadstone, the number of sources of gritstone declined. Now that there is an increasing need for highly polish-resistant aggregates, a number of these old quarries have been modernised and re-opened and other sources are being sought. The cost of production is greater than with other rocks because of their abrasive nature (the very characteristic that singles them out as polish-resistant stones), but the demand is such as to make them economically viable again.

The most important sources of gritstone aggregates in England have been in Yorkshire, Lancashire and Devon. Gritstone formations are also extensively quarried in Scotland and Wales.

b.   Use of gritstone in road surfacings

The gritstone group of rocks includes some very weak sandstones that are unsuitable for use in roadmaking. This has often led to engineers discriminating against the whole group. An examination was made of the 85 samples of gritstone from all over the country (Director of Road Research, 1954). On assessment of the behaviour of bituminous surfacings made with these different types of gritstone, it was considered that 64 were suitable for use as roadstone and 21 were not. All the samples were subsequently examined by a petrologist, who found that the unsuitable gritstones were generally coarse-grained sandstones deficient in "cement", whilst the accepted ones were either greywackes, tuffs, breccias, fine-grained well-cemented sandstones, flagstones or siltstones.

c.   Gritstone survey

A survey by the Transport and Road Research Laboratory in collaboration with the Institute of Geological Sciences was carried out to establish British resources of arenaceous rocks (sandstones and similar rocks) with the object of finding new sources of aggregate for skid-resistant surfacings (Hawkes and Hosking, 1972). It was concluded that there are large resources of high-quality material available; also, information was provided that will be of value in future searches for similar material and in selecting stone within a quarry. Results showed that the desirable qualities (high resistance to polishing and abrasion) in arenaceous rocks are more dependent on their geological history than on their mineral composition. Further information on this survey is given in 2.1.6 below.

2.1.6 Polish-resistant aggregates

2.1.6.1 Natural roadstone

Knill was the first to make a comprehensive attempt to relate polished-stone value (PSV) to the mineralogical and textural parameters that characterise rocks in different roadstone groups (the group classification of BS 812:1967) (Knill, 1960). In all 76 samples belonging to nine of the ten groups of naturally-occurring materials were examined, but because they represented a comparatively small sample of each group, detailed correlations were not possible. Nevertheless she was able to show that the polishing resistance of a rock is not determined specifically by its mineralogical or chemical composition. Also, it was found that that groups of very different mineralogical composition showed a remarkably similar distribution between rather poor and good polishing characteristics. An important conclusion was that the resistance to polishing is influenced as much by the degree of bonding between mineral constituents as by any other single factor. Bonding characteristics are complex and reflect the composition and origin of a rock and its subsequent history.

Knill's work showed that the gritstone group provided most of the aggregates with good to excellent polishing resistance.

Examination of Knill's results coupled with general testing experience led the researchers (Hawkes and Hosking, 1972) to the view that the testing of all British rocks to be too time-consuming and would only tend to confirm present knowledge rather than add to it. Research was therefore directed towards the arenaceous rocks, which comprise a variety of individual rock types classified chiefly in the gritstone group, with some representatives in the quartzite and schist groups included in the programme. This choice of direction was not meant to imply that rocks in the other groups would always prove unsatisfactory, only that the arenaceous rocks seemed to be the most promising source of consistently highly polish-resistant material. An additional factor was that these types of rocks had been exploited to a lesser extent during the last few decades, probably because they are less suited to the quarrying techniques of the times and the emphasis on the desirability of high strength for an aggregate.

A list of some 151 potential sources was drawn up on the basis of geological age and variations that occur in the different parts of Great Britain. These locations were visited and after a provisional examination for adequate aggregate strength, samples of about 50 kg were collected for subsequent detailed examination and testing. In the event the rocks at 65 sites were found to be either too weak for road-making or too thinly interbedded with slate to permit economic working. Samples from the remaining 86 sites were subjected to detailed examination.

The aggregates tests employed were:

i   Aggregate abrasion value test.
ii   Aggregate impact value test (the modified test was used for the weaker samples).
iii   Polished-stone value test.

Other tests included:

i   Ultrasonic wave velocity determinations on water-saturated and dried samples.

ii   Density of water-saturated and dried samples.

Petrographic examination included the recording of the following:

i   Size of largest fragments and average grain size of all constituents.
ii   Total percentages of fragments and matrix.
iii   Individual percentages of the following components irrespective of size: quartz, feldspar, calcite, clay minerals, lithic fragments and any other minerals found in appreciable quantities.

Five of the 86 samples were found to combine an AAV of 10 or less with a PSV of 70 or more, and a further 18 samples were found to have a PSV within the range 65-69. A further 17 samples had PSVs greater than 60.

The main findings of the survey were:

i   Great Britain has potentially large resources of high-quality road-surfacing aggregates. Much of this material would meet more stringent requirements than were currently being implemented.

ii   The best materials came from Wales; very good material was found in Northern Ireland, southern Scotland, and the English counties of Shropshire, Devon and Cornwall.

iii   The petrographic studies indicated that the degree of consolidation of an arenaceous rock is more important than composition and grain size in determining the specific values of mechanical and physical properties.

2.1.6.2 Calcined bauxites

Bauxite (a rock consisting largely of hydrated aluminium oxides) is best known as an ore of aluminium. However certain bauxites are used to manufacture refractory and abrasive materials by calcination at high temperatures. Some of these calcined bauxites are of value as polish-resistant road aggregates.

a.   *Use in conventional surfacings*

Three full-scale road experiments included sections of surface dressings using calcined bauxite aggregate (a refractory grade known as "RASC grade"). All sections showed an exceptionally high resistance to skidding. Details are given in Chapter 5.5.

The calcined bauxite chippings for these experiments involved the screening of large quantities of material. RASC grade calcined bauxite is manufactured for refractory purposes and, for this reason, is quenched in water. The result is a shattered material mostly too small in size to be used in conventional surfacings. Unless calcined bauxite is specially manufactured for roadmaking purposes it is unlikely that it will be available in sufficient quantity to be of importance in conventional surfacings.

b.   Use in resin-bound skid-resistant surfacings

High-quality refractory-grade calcined bauxite from Guyana, when used as a 3 mm grit in resin based binders, was found to maintain an exceptionally high resistance to skidding under the most severe traffic conditions (Road Research Laboratory, 1970). However, material from this source was expensive and, because of the demands of the steel industry, it was in very short supply. Work was therefore been undertaken with the object of finding other calcined bauxites that would serve the same purpose.

As part of this research, seven samples of high-temperature calcined bauxite from Guyana of different chemical and mineralogical compositions were selected from a batch of refractory-grade bauxite. These were examined and tested in the laboratory and then tried under actual road conditions at two heavily trafficked sites. Other calcined bauxites and other materials were also studied in a similar way; results showed that the best performance was achieved with materials containing a high proportion of small strongly bound alpha-alumina (corundum) crystals.

A further study (Tubey and Hosking, 1972) showed that abrasion- and polish-resistant properties were associated with materials containing numerous small (15 μm to 70 μm) corundum crystals bonded by a moderate quantity of glassy "cement" to give an open-textured granular surface. Chemical analyses of these materials showed a high alumina content, a low silica content, a fairly low iron content and a negligible loss on ignition. The corresponding mineral analyses showed abundant corundum, little or no mullite, and no significant quantities of other crystalline materials.

This work showed that other bauxites, if calcined under appropriate conditions, could have equally satisfactory properties. Bauxites from Australia, Ghana, and Northern Ireland were therefore calcined in the Laboratory's rotary furnace at temperatures up to 1600 degrees C. These trials (Hosking and Tubey, 1973) showed that one (from Ghana) was as good as RASC grade bauxite, another (from Australia) was nearly as good and a third (from Northern Ireland), although poorer than the others, was better than any natural roadstone. This work also indicated that a suitable minimum alumina content would be 65 percent and that the silica content should be less than 10 percent. Moreover it also showed that a fairly high iron content was not detrimental, indeed the presence of iron was found to improve the product and reduce the temperature needed for calcination.

2.1.6.3 Synthetic aggregates of high polish-resistance

High resistance to polishing is generally associated with low abrasion resistance but the latter property is important at heavily trafficked sites. Such aggregates are scarce and high haulage costs make them very costly in some regions of the country. The highest polished-stone value (PSV) for a natural roadstone suitable for use in surfacings is about 70, but the abrasion resistance of such stone is only marginally acceptable.

Calcined bauxite has a very high PSV (75) and is very durable but it is very expensive and in short supply. A search for a cheaper substitute was made in the mid-1960s but, although synthetic roadstones with even higher PSVs were investigated, they were either more costly or lacked adequate abrasion resistance.

Research into the manufacture of suitable aggregates has been carried out by several organizations and a review published (Hosking, 1976). A number of types of polish-resistant roadstone, first defined by James (1968), were studied: an outline of the more important is given below:

a.   Hard aggregates

Very hard materials which are virtually unaffected by the abrasive and polishing action of tyres. Such materials (diamond, corundum, etc.) are very costly and calcined bauxite is the least expensive that has to date proved satisfactory (see 2.1.6.2 above).

b.   Conglomerations

Conglomerations of small hard particles which maintain an unpolished surface by virtue of the progressive removal of individual particles. The most polish-resistant roadstones (sandstones and gritstones) are of this type, but in general they tend to wear away rather rapidly (see 2.1.6.1 above).

c.   Gritty aggregates

Dispersions of hard particles in a softer matrix. This principle was first suggested by the properties of gritty limestones, where quite small admixtures of grit can profoundly affect the polishing resistance of the stone.

A study (Hosking, 1970) has been made of this class of aggregates that owe their polish-resistant properties to the differential wear of hard grit particles dispersed in a softer matrix. Fig. 1 shows

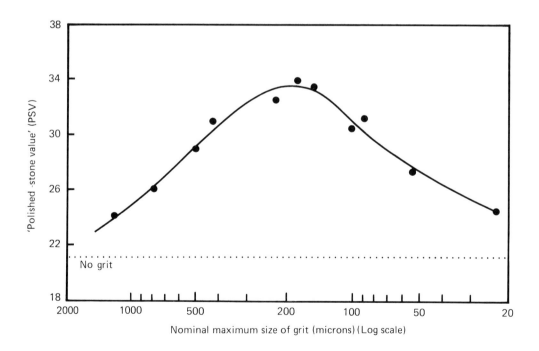

Fig. 1   The relation between size of grit and the "PSV" of moulded resin specimens.

the relation that was established between the grading of the grit and polish resistance; best results were obtained with a grit of 300 μm-150 μm in size. The effect of the proportion of grit is shown in Fig. 2, the benefit of additions above 20 per cent was small and the ensuing product tended to lack abrasion resistance. Study of the effect of different types of grit showed that calcined bauxite gave the best results and the more expensive silicon carbide was nearly as good. "Aloxite" (a commercial abrasive), crushed flint, Chertsey sand and emery also gave good results and were cheaper. A range of matrices were found to show promise including fluxed Etruria marl and a fluxed white-firing ball clay.

The result of this research was used to draw up a Patent Specification, which was filed at the Patent Office on 22nd August 1967 (James and Hosking, 1967).

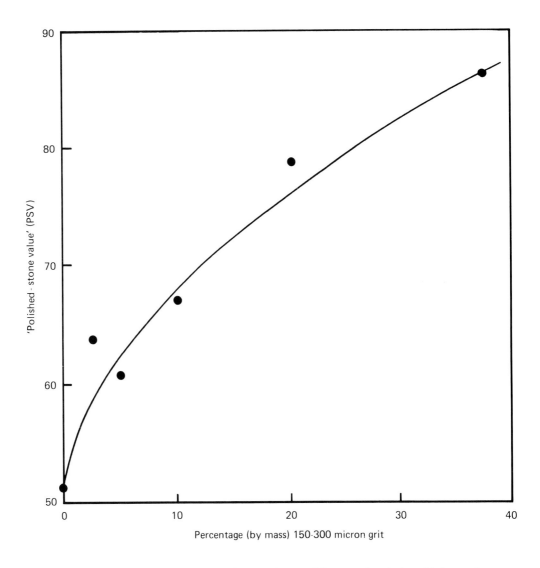

Fig. 2   *Relation between the PSV of an artificial roadstone and the proportion of grit used in its manufacture.*

d.   Brittle aggregates

Materials which fracture under traffic in an irregular manner and so maintain a rough surface, for example, calcined flint.

e.   Vesicular aggregates

Vesicular materials such as some blastfurnace slags and pumice have a greater resistance to polishing than when in a dense form. The lower density of such materials (provided that they are sufficiently strong and durable) is an advantage in that they are comparatively cheaper to use when purchased by weight.

Research has indicated (Hosking, 1974) that increasing the porosity of a synthetic roadstone resulted in better polishing resistance but poorer abrasion resistance. Optimum properties were obtained with a class of ceramic roadstones when the porosity was increased to between 25 and 35 per cent by the addition of combustible dust of 0.125 mm nominal-size before firing.

f.   Blastfurnace-slag based aggregates

Work described by Sweetman (Sweetman, 1974) was carried out into the pilot-scale manufacture of an aggregate using particles of calcined bauxite in a blastfurnace slag as an matrix. The investigation showed that the process is not likely to be commercially viable because it would require rapid mixing of the ingredients at a high temperature followed by rapid cooling, which would be incompatible with the needs of commercial production.

g.   Shaped aggregates

A disadvantage of the manufacture of synthetic aggregates is that during the crushing and screening processes much fine material is produced. This can amount to as much as one half of the material crushed, which would either have to be recycled or sold at a loss. A way of overcoming this difficulty would be to manufacture shaped aggregate particles that could be used without further processing. This would also improve grading and eliminate dust. With this end in view studies were made (Hosking and Jacobs, 1974) of the behaviour of a number of shaped aggregates (spheres, cylinders and prisms) when used as chippings in rolled asphalt. Of the shapes studied, cylinders and hexagonal prisms gave the best texture, but additional texture would be necessary in order to provide acceptable skid-resistance.

2.2   Production

The production of aggregate consists of two processes, extraction and processing. The method of extraction will depend upon the type of material and its location. At one extreme extraction consists simply of digging a material such as gravel from the ground and, at the other, blasting solid rock. Other extraction processes include dredging and mining. Most deposits are covered by an overburden of lower-grade weathered material that has to be removed before extraction begins. This may be replaced after extraction or used for less demanding engineering purposes.

After the raw material has been extracted from the quarry, gravel pit or other source, it is usually processed by crushing and screening to produce a range of "single-sized" aggregates. (*Author's Note*: In the most recent revision of BS 63 the term "single-size" has been employed. The Author has retained the earlier expression because it is that used in virtually all of the quoted publications.) Many deposits include pockets of lower-grade material that have to be removed before the marketable product can be reached. "Scalping" is a process commonly employed for this purpose. This involves the screening off of the smaller size material after the first stage of crushing (primary crushing).

Extraction and processing have been described in detail in publications such as the Engineering Geology Special Publication No.1 (Geological Society, 1985).

The later stages of crushing (secondary and tertiary crushing) and screening can have a considerable affect on the final product. Crushing affects the shape of the product as well as the relative proportions of the different sizes produced. Screening affects the grading of the single-sized products. Both these processes have been studied by the Road Research Laboratory; an outline of this work is given in 2.2.1 and 2.2.2 below.

2.2.1 Crushing

By 1937 the Laboratory had commenced a study of the effect of crushing methods on the shape of chippings. This initial work was followed by the purchase of a full-scale crushing and screening plant in 1950 in order to study the characteristics of crushers; it was installed in a Somerset County Council's quarry near Wells.

Early experimental work with the crushing plant showed that, compared with machine crushing, a higher proportion of single-sized material could be obtained by hand breaking, but the shape was no better (Director of Road Research, 1954).

The results of the completed study made with the plant at Wells were published as a Technical Paper (Shergold, 1959). This was an extensive investigation of the performance of the crushers used in the final stage of the crushing of roadmaking aggregates; these crushers are known as "granulators". The plant was designed so that the operation could be studied under a range of controlled conditions. During 4 1/2 years of continuous operation, the behaviour of ten granulators were studied. These included single- and double-toggle jaw granulators, cone granulators, crushing rolls and fixed- and swing-hammer impact breakers. Each machine was studied using a number of different types of rock (12 in all) and over a wide range of operating conditions. Factors studied included the type of rock, the grading, shape and moisture content of the feed to the crusher, and the rate of feed, the crusher speed and setting. Special characteristics such as the length of stroke of the jaw granulators, fine and coarse linings in the cone granulators, plain and serrated roll shells in crushing rolls and the use of a grid in the outlet of a swing-hammer breaker were also investigated. More than 500 test runs were made and on each run determinations were made of the grading of the feed, the grading and shape of the product, the output in tonnes per hour, the horsepower developed by the motor and the energy consumption in kilowatt hours per tonne of rock crushed.

The main object of the work was to assist producers of aggregates to meet fluctuating demands for aggregates and to keep within specification limits for grading and shape. Detailed information is given in the Technical Paper, the main conclusions of general interest are summarised below.

The factors affecting the grading of the product were found to be as follows:

i   The main factor was the setting of the jaw granulators, cone granulators and the crushing rolls. For the impact breakers the factors were the speed of the hammers and the size of the grid (if any) in the outlet. Varying these factors was the normal way of controlling the size of the product.

ii With the jaw and cone granulator and crushing rolls, the predominant size of the chipping in the product approximated to the closed setting but, because other factors affect the grading of the product, the exact setting required to produce the maximum proportion of a given size of chipping had to be determined by trial and error.

iii The highest proportion of the three principal sizes of chippings that could be obtained in the product under normal crushing conditions are shown in Table 1.

The fines from the impact breaker were found to be coarser than those from the other machines and, in all the machines, crushing to produce the highest possible proportion of 3/8 inch (9.5 mm) chippings resulted in a very high proportion of fines.

iv The proportions (see Table 1) could be increased slightly by closed-circuit crushing or by a reduction in the rate of feed, but a reduction in total output resulted. In jaw granulators, the proportions were lower when jaws were worn.

v   In jaw and cone granulators, increasing the speed of the machine resulted in a finer product and a disproportionate increase in fines. A reduction in the rate of feed resulted in a decrease in the proportion of fines.

vi Other factors, such as the misalignment of the jaws, the wear of the hammers, the use of special fine concaves in cone granulators, and the use of serrated shells on crushing rolls, affected the fineness of the product but had little or no effect on the proportions in which the different sizes of chipping were produced.

vii The grading of the product was only slightly affected by the grading or shape of the feed except in the crushing rolls, where a larger feed gave a finer product.

TABLE 1
Highest proportions of the three principal sizes of chippings obtained under normal crushing conditions

	Chipping size		
	3/4 in (19 mm)	1/2 in (14 mm)	3/8 in (9.5 mm)
Crusher	(percentage produced)		
Jaw, cone, or roll granulators	40	20	30
Impact breakers, restricted outlets	27	16	20
Impact breakers, un-restricted outlets	20	12	18

viii The strongest rocks required 1/4 inch (6.3 mm) closer setting with jaw granulators, 1/8 inch (3.2 mm) with cone granulators, or 100 rev/min higher speed with an 18 inch (460 mm) impact breaker, to obtain as fine a product as with the weakest rocks. The fines from all the rocks had a lower proportion of the No.25 to No.100 (600 µm to 150 µm) sieve size fraction than is found with natural sands.

The factors that were found to affect the particle shape of the product were as follows:

i The dominant factor was found to be the reduction ratio (ratio of size of feed to the size of product).

ii The effect of the type of granulator was less than the effect of reduction ratio. Impact breakers were best, followed by jaw granulators, crushing rolls and cone granulators.

iii Some types of rock tended to give better-shaped products than others. In general, stronger and finer-grained rocks tended to give a poorer-shaped product.

iv Slightly poorer shape was given by crushers with slower rates of feed (except impact breakers), worn hammers, dry feeds, cubical feeds and open-circuit crushing conditions.

The methods that were developed for assessing the performance of rock granulators under controlled conditions on a test were published in Road Note No. 37 (Road Research Laboratory and Military Engineering Experimental Establishment, 1965).

2.2.2 Screening

Research into methods of measuring of the efficiency of screening by the Road Research Laboratory began in 1954, together with studies of the single-deck vibratory screen installed at the Laboratory.

It was found that the most reasonable values for the efficiency of screening were obtained by subtracting from 100 the percentages by weight of the "difficult" particles that were incorrectly classified (i.e. oversize passed or undersize retained). "Difficult" sizes were defined as those immediately larger or smaller than the desired size of separation, in the series .....28 mm, 20 mm, 14 mm, 10 mm, ... The resulting formula is:

$$E = 100 - (HF/B) - (JG/A)$$

where

E = efficiency of screening or separation
A = percentage of difficult oversize in the feed
B = percentage of difficult undersize in the feed
H = percentage of difficult undersize on the material retained by the screen
J = percentage of difficult oversize in the material passed by the screen
F = percentage of the feed retained by the screen
G = percentage of the feed passed by the screen

An efficiency of 70 to 80 per cent was found (Hosking, 1964) to be necessary to produce chippings to the standards of BS 63, the British Standard for single-sized aggregates.

Investigations had been extended to working quarries by 1964 and methods were developed for measuring the efficiency and other characteristics of screening plant under normal operating conditions and without interfering with production. It was found that the efficiency of separation of the screens could be calculated from the grading of the single-sized chippings passed to the bins and from the quantities passing the individual screens. The amplitude, frequency and direction of vibration of the screens was measured by means of a stroboscope and a specially designed chart, and the angle of slope was measured with an inclinometer. Special taper gauges were designed to measure accurately the aperture sizes of the screens. These methods were found to be of direct use to the operator of the quarry in which they were developed, helping to locate causes of incorrect grading and deficiencies in the proportion of certain sizes of aggregate. They were also found to be of use in forecasting the effect of changes in the plant on the grading of the screened products. Full descriptions of these methods have been published (Hosking, 1965).

2.2.3 Distribution

2.2.3.1 General

The quantities and prices of materials used in new road construction were analysed (James, 1972) as a preliminary to programmes of research by the Transport and Road Research Laboratory on the economics and transport of road materials. These covered both conventional materials and waste.

2.2.3.2 Waste materials and transport

A study (Transport and Road Research Laboratory, 1973) showed that about 250 million tonnes of crushed rock, gravel, sand, etc., were being produced in the United Kingdom each year. Of this about one-third was used in roads, one-third in buildings and one-third for miscellaneous purposes. At the same time over 100 million tonnes of solid mineral wastes were being produced as by-products of industries such as mining and metallurgy, and less than one-third of this was being put to use. Much research had been carried out in the preceding decade on the technology of using wastes in roads, and the engineering problems had been largely identified and solved. The remaining problems that were were found to be inhibiting more extensive use of wastes in road construction were primarily economic and environmental. To move waste from source to site was usually found to be more costly than to dig local earth fill or aggregates.

2.2.3.3 Aggregates distribution

Studies of the distribution of aggregates have been published (Glover, 1976; Transport and Road Research Laboratory, 1977). A survey of waterways and coastal shipping provided data on the "track" represented by commercial inland waterways and coastal shipping routes, and on the operating costs of the vessels using them. The calculated operating costs of barges and coastal vessels at the time were found to range from 2p to 6p per tonne-mile (depending on the size of the vessel) for aggregates and other bulk cargoes. These costs excluded those for handling facilities such as cranes, conveyers and wharfs, which were considered separately.

TABLE 2
Cost of transport of aggregate by different means

Means of transport	Pence/tonne-mile at 1975/6 price levels
Lorry (15 tonnes capacity)	4
Train (650 tonnes capacity)	0.6
Barge (560 tonnes capacity)	2
Barge (100 tonnes capacity)	6
Coastal vessel (2500 tonnes capacity)	0.5

By including this information into the model of aggregates distribution it was found possible to make broad comparisons between the operating costs (per tonne-mile) of a trip exceeding about 30 miles by lorry, train and loaded barge and coaster.

Table 2 gives estimates of the cost of transport by the different means assuming that there were suitable paths between the origin and destination of the trip. Those costs applied to a fully loaded lorry, train or vessel moving at a constant speed and it was assumed that the return journey was made empty.

A further study (Glover and Shane, 1979) was made using a model that was a balanced representation of all the interactions between the different processes necessary to produce, handle, transport and finally prepare aggregates at the sites where they were used. Distribution was represented at regional level for seven aggregates and eight end-uses and the model included alternative means of distribution between regions; it was therefore possible to investigate possible situations at the national and regional levels. It took account of transport within regions but assumed that it was entirely by road.

The annual demand for aggregates in Great Britain was about 230 million tonnes and in most parts of the country was being met mainly from local sources. However, constraints imposed by declining reserves and planning control were leading to deficits in some areas, notably in the South East and the Manchester/Merseyside regions. The model was developed as a tool to assist planners to evaluate possible policies for extracting and transporting aggregates and was used for a number of investigations including:

i   The effects of shortages of materials in some regions and the possible imports of materials from other regions by road and rail.

ii The effects of changes in the density of rail receiving depots.

iii The scope for transporting materials by sea from coastal super-quarries.

iv The investigation of distribution patterns which minimise the environmental disturbance of transporting aggregates.

Some of the conclusions of the study were:

i   The total identifiable costs of the distribution of aggregates were about £630M, at 1974 price levels and at 1972 levels of demand.

ii In resource cost terms, aggregates could be transported more economically by rail (with final deliveries by road) than by road only, provided that the origin-destination distances were at least 25 miles and the road delivery journeys were short compared with rail distances. This conclusion, however, depended on rail depots, locomotives and rolling stock being available at defined levels of cost. If rail depots become more sparse, road transport increased as the expense of rail and rail transport became uneconomic if the final road delivery distance was greater than about 45 miles.

iii At the current price levels, there was little or no market for general purpose aggregate moved by sea from coastal quarries in Cornwall or Scotland unless there were severe shortages of sharp sand and gravel in the South East and of limestone and hard rock in the Midlands and South West. Less severe shortages were made up more economically by rail movement of granitic rock from Cornwall. As land-won aggregates became more and more scarce, a point was reached where movements of stone by sea become economically feasible first from Cornwall and then from Galloway, provided that rail capacity, sea transport costs or extraction costs at coastal sites were reduced considerably.

iv Because nearly all aggregates have to be delivered to their final destination by road, the total road traffic (expressed in tonne-miles) was not greatly reduced by substitution of rail or coastal vessel for the main haul.

"Guidelines for aggregate provision in England and Wales" (DOE, 1989) reviews the distribution of aggregates and gives advice to mineral planning authorities and the industry aimed at ensuring that the industry continues to receive an adequate and steady supply of aggregates at the best balance of social, environmental and economic costs.

2.3 Waste and low-grade aggregates

2.3.1 General

It is in the national interest to make use of waste and low-grade materials as alternatives to exploiting reserves of naturally occurring natural aggregate as much as is economically possible. It would also assist in the disposal of unwanted waste materials. Road construction could provide a major outlet for the utilisation of such materials.

The Transport and Road Research Laboratory reviewed the situation (Sherwood, 1974) and subsequently made separate studies of road potential of the sources of waste and low-grade materials in Great Britain. Table 3 gives a summary of the quantities of major wastes available and the results of the studies of individual wastes are outlined in the following sections of this chapter.

TABLE 3
Quantities of major waste materials available

Material	Total quantity	Annual production	Annual use
		(quantities in million tonnes)	
Colliery shale	3000	50	8
Pulverised fuel ash	-	7.5	6
Furnace bottom ash	-	2.5	6
China clay waste (sand)	125	10	1
(other wastes)	155	12	1
Slate waste	300-500	1.2	0.03
Incinerator waste	-	0.6	-
Hassock	?	?	?
Spent oil shale	330	None	?

Studies showed that approximately 24 million tonnes of natural aggregates were being used for road surfacing annually, and that the corresponding figures for road base, sub-base and imported granular fill were 16 million, 16 million and 3 million tonnes respectively. It was concluded that although the use of waste materials in road making can play a part in reducing the accumulation of waste tips, it has a relatively minor role. However a more encouraging conclusion was that their use would be of value in conserving good-quality material that might otherwise be used.

Possible ways in which waste materials might be used in roads are as fill material and in base and sub-base material in new road construction schemes. The latter materials would be Types 1 and 2 granular sub-base material, soil cement and cement-bound granular material.

A recent publication (Arup, 1991) gives a review of the occurrence and utilisation of minerals and construction waste.

2.3.2 Colliery shale

2.3.2.1 General

Colliery shale deposits are composed of the waste products from coal mining which are either removed to gain and maintain access to the coal faces or unavoidably brought out of the pit with the coal. They are usually dumped together with small quantities of coal and other washings. This gives rise to considerable variation which is further increased by spontaneous combustion which occurs in some of the heaps.

Colliery shale mainly occurs in the Midlands, the North of England, South Wales and Central Scotland.

2.3.2.2 A study of unburnt colliery shale

Following the disaster at Aberfan, public attention was focused on methods of getting rid of the waste heaps that exist close to the mine workings. An obvious way was to use these materials for road works. However unburnt shales had been used only in small quantities and, until the mid-1960s, no nationally agreed policy had been formulated. The Road Research Laboratory surveyed those properties (Road Research Laboratory, 1968) of this material that affect road work and had discussions with National Coal Board (NCB) representatives and engineers in the Road Construction Units, Scottish Development Department and Welsh Office. From this assessment it became clear that the possibility of spontaneous combustion was the one factor that caused many engineers to be uncertain about using unburnt colliery shale as fill in road works. A Technical Note was prepared outlining the factors to be considered in using unburnt colliery shale as fill in road works, and this suggested that the material could be used under guidance from the NCB. The NCB welcomed this and formed a small unit to advise engineers on the use of waste materials from the coalfields. A Technical Memorandum (Ministry of Transport, 1968) was issued to provide, for the first time, a nationally agreed policy on the use of these waste materials in road works.

2.3.2.3 A review of usage in 1975

A review of the usage and potential of colliery shale in road construction has been made by Sherwood (Sherwood, 1975a). It was established that about 50 million tonnes were being produced each year, that 3000 million tonnes had already accumulated in spoil heaps and that about 6 million tonnes were already being used, mostly in roadmaking.

The major outlet was as a fill material where the limiting factors were a fear of spontaneous heating (no longer considered a problem), excessive sulphate content (which can leach out to damage concrete and cement-bound materials) and frost-susceptibility. Limits were therefore imposed for sulphate content where it could damage cement-bound layers or bridge abutments. Most unburnt shales and some burnt shales are frost susceptible; in such cases care must be taken to ensure at least 450 mm of cover.

Similar considerations apply to the use of shale as a granular sub-base. However cement stabilisation can be used to reduce frost susceptibility and to allow the use of shales lying outside the grading limits for granular sub-bases.

Virtually all burnt shales and most unburnt shales may also be used as base materials when stabilised with cement.

Another use of colliery shales is in the production of artificial aggregates. These are of lower density than naturally occurring aggregates and have more value in building construction than in roadmaking. The quantity used in this way does not make a significant impact on the amount of shale used in road works.

2.3.3 Pulverised fuel ash

The properties and potential use of pulverised fuel ash in roadmaking were also investigated by Sherwood (Sherwood, 1975b).

Most coal-burning power stations in this country burn coal which has been pulverised into a fine powder. This coal produces a very fine ash which is carried out of the furnace with the flue gases.

This ash is known as pulverised fuel ash (PFA) and accounts for about 75 to 85 per cent of the ash formed from the burnt coal. The remaining coarser fraction of the ash falls to the bottom of the furnace where it sinters to form a coarse material known as furnace bottom ash (FBA). PFA is precipitated from the flue gases and is supplied as a dry powder (hopper ash) or has water added to form "conditioned PFA", which can be stockpiled or used immediately. "Lagoon PFA" is a form recovered from storage in lagoons and is often mixed with the coarser FBA.

PFA, sometimes referred to as "fly ash", resembles Portland cement in appearance and usually has some pozzolanic (cementitious) properties. The coal-burning power stations are mostly near the coalfields, which mean that the major sources of residue are Central England (approximately 6.8 million tonnes per annum), London and South-East England (1.4 million tonnes), North-East England (0.9 million tonnes) and South Wales & Bristol (0.8 million tonnes). It has the advantage over some waste materials in that it is produced in large uniform and predictable quantities.

PFA can be used at all levels in the road structure, but the scope for using it in road construction in appreciable quantities above sub-base level is severely restricted. The major outlet is as a fill material, where it is used in large quantities. Its self-cementing properties make it of particular value where settlement problems are possible, such as behind bridge abutments. It should be borne in mind that it is frost susceptible and therefore should not be used on its own within 450 mm of the surface.

PFA is of value as a sub-base material when stabilised with cement or hydrated lime; its own self-cementing properties alone are not sufficient for this purpose. As a road-base material it is only suitable for lightly trafficked roads. However in concrete it can be used in three ways:

a. *PFA as a constituent of concrete*

A possible advantage of adding PFA to concrete is to reduce the water content necessary and improve workability. For some applications, such as mass-concrete dam construction, the inclusion of selected PFA in concrete provides a specific technical benefit in the form of reduced heat generation.

b. *PFA as a component of cement*

It can be ground with cement clinker.

c. *PFA as a raw material in the manufacture of cement*

It is added at the pre-kiln stage.

2.3.4 Incinerated domestic refuse

The Transport and Road Research Laboratory's Report LR 728 (Roe, 1976) describes an investigation of the properties of incinerated domestic refuse.

Direct incineration, which reduces the volume of the refuse by about 90 per cent, has been used increasingly in recent years to reduce demand for tipping space. There are now some 33 large incinerators in England which between them produce about 1 million tonnes of residue per year within the major urban areas.

Incinerated refuse consists mainly of clinker and glass with small amounts of unburnt matter such as paper and garden refuse. Most of the ferrous metal is usually extracted at the plant but some metals remain in the residue, the amounts varying from plant to plant.

The study has shown that incinerated domestic refuse is a material having considerable roadmaking potential. Ashes differ considerably from plant to plant but individual plants usually produce reasonably consistent material. Most ashes are suitable for use as bulk fill and some may be suitable for use as Type 2 granular sub-base. Incinerated wastes may have a fairly high sulphate content, in which case they should not be used near concrete structures or cement-bound materials. As the material is produced within urban areas, its use in urban road construction has obvious advantages.

2.3.5 China clay sand

China clay sand was studied as part of an investigation on the utilisation of waste materials in road construction (Tubey, 1978).

For each tonne of china clay extracted, approximately nine tonnes of waste arises of which 3 1/2 tonnes is sand. 10 million tonnes are produced annually and there is about 125 million tonnes present in waste heaps. China clay sand is predominantly composed of quartz grains, but also includes small proportions of other materials such as mica.

The samples of sand studied had gradings that conformed to the requirements for granular sub-base materials and have been used for this purpose in Cornwall. They could be stabilised with cement to meet all the requirements of cement-bound granular road-base construction. They were also suitable for use in concrete and, with treatment, could also be used for sand asphalt and rolled asphalt. Synthetic aggregates with a high resistance to polishing can also be made from china clay sand.

It would seem that china clay sand has a considerable potential as a roadmaking material. Unfortunately little road construction is planned in the area where it occurs (Devon and South-East Cornwall) and the high cost of transport to areas where the demand exists limits its usage.

2.3.6 Spent oil shale

Spent oil shale was included in the Transport and Road Research Laboratory's study of available information on major waste products in relation to their suitability for roadmaking (Burns, 1978).

During the production of oil from oil shale the spent shale has been deposited in large heaps (bings) on land beside the mines and refineries. Approximately 330 million tonnes occur in the West Lothian district of Scotland.

Tests showed that the material complied with the grading and strength requirements for sub-bases, but that a relatively high sulphate content precludes its use near concrete structures. Frost heave tests showed that there could be problems with this material, but this can be reduced to an acceptable level by the addition of small amounts of cement. By 1978 about 25 million tonnes had been used as bulk fill, between 0.5 and 1 million tonnes as upper sub-base material in cement stabilised form and another 1 million tonnes in the lower sub-base. A proposed use outside roadmaking is in landscaping schemes.

2.3.7   Miscellaneous wastes

2.3.7.1 General

A report by Sherwood (Sherwood *et al*, 1977) deals with a number of wastes which for one reason or another have less potential than those described above; they are cement-kiln dust, slate waste, demolition wastes, non-ferrous slags, waste rubber and waste glass.

2.3.7.2 Slate waste

Slate waste is the only one of these materials that occurs in comparable quantities (500 million tonnes) to the materials considered above. It occurs in large quantities in North Wales, the Lake District, Devon and Cornwall. However it is a declining industry and only about 1.2 million tonnes per year was being produced in 1977.

The flaky nature of the particles cause problems in compaction but it has been used for embankment and sub-base construction in North Wales. The fact that it is produced in remote areas of the country severely limits its more widespread use.

2.3.7.3 Cement-kiln dust

This is a product from the manufacture of cement and is a very fine powder resembling cement in appearance. About 0.4 million tonnes are produced annually and there was a stockpile of about 3 million tonnes in 1977. Freshly produced cement-kiln dust has little application in roadmaking, but well-weathered material can be used in earthworks; it is a relatively light-weight material and has some residual cementitious properties.

2.3.7.4 Demolition wastes

Much of the material produced from demolition is unsuitable for roadmaking because it becomes hopelessly mixed with the minor components such as plaster, wood and glass. However some demolition wastes such as brick and concrete rubble have high potential for sub-base construction; the major problem in their use is their spasmodic occurrence which makes their planned use in road works difficult.

2.3.7.5 Metallurgical (non-ferrous) slags

These are produced from the smelting of non-ferrous ores. The amount of non-ferrous slags produced in this country is small (250,000 tonnes/year). The slags may have potential uses in the vicinity of their origin, but care needs to be taken to ensure that they are stable in the presence of water and they do not contain components that might give rise to water pollution. (See also 2.1.1.7.)

2.3.7.6 Waste glass

There is potentially 1.5 million tonnes/year of waste glass available for use in road construction. This compares favourably with the amounts of other waste materials already being used in roadmaking. Unfortunately this waste is widely dispersed, making collection difficult and only justifiable where a high-grade use can be found. The only outlet for it in road construction appears to be as an aggregate in bituminous mixtures, but the skid-resistance properties of glass are markedly inferior to those of other more readily available materials.

2.3.7.7 Waste rubber

There is about 75,000 tonnes/year of waste rubber available for use in roadmaking which, at present, finds no commercial use and which presents problems of disposal. Its use in road construction is likely to remain limited and will not significantly reduce the disposal problems of waste tyres, or the very small amount (about 50 tonnes/year) of natural rubber used in roads to improve the properties of bituminous binders. This is because although waste rubber would be cheaper, the price advantage tends to be cancelled out by difficulties in blending it with bitumen. Nevertheless some interest is currently taken in the utilisation of waste rubber in road making, particularly in Europe.

2.3.7.8 Furnace clinker

Furnace clinker is the residue from burning lump coal at the older power stations. It can be used as a concrete aggregate (BS 1165:1966) but is unlikely to be available for roadmaking in any appreciable quantity.

2.3.7.9 Quarry wastes

Most quarrying operations produce waste and low-grade materials that can be used in road construction. Of these china clay sand and slate waste have been reviewed above. Hassock is another such material. In the area around Maidstone in Kent, a fairly hard sandy limestone known as "ragstone" has been quarried for many years. It occurs, however, in deposits that are interbedded with layers of soft calcareous sand or argillaceous sandstone known as "hassock" which forms as much as 80 per cent of the quarried material. This material can be used as fill and, when mixed with cement, for base and sub-base construction.

2.3.7.10 Chalk

Although chalk is not a waste product, it is a low-grade material frequently used in roadmaking. It represents 15 per cent of the major geological formations of England. The main characteristic that precludes it from more general roadmaking purposes is its frost susceptibility. However it can be used as fill material and when stabilised with cement as a base and sub-base.

2.3.7.11 Blastfurnace and steel slags

These are extensively used in roadmaking and are the subject of Chapter 2.1.1 of this book.

3 General requirements of roadmaking aggregates

3.1 Practice in Great Britain

3.1.1 Water-bound and unbound materials

3.1.1.1 Water-bound macadam

The properties of aggregates were first studied in relation to water-bound roads, and although water-bound macadam is now not much used because of traffic levels, some reference should be made to the properties desirable in the aggregate to be used for this purpose.

Water-bound macadam depended for stability upon the mechanical interlocking of the particles, the fine material required for binding being added only after the maximum interlocking had been obtained by rolling the dry stone. It was essential, therefore, that the stone should have a high crushing strength.

For similar reasons the resistance to attrition, abrasion and impact needed to be as high as possible in the days of horse drawn vehicles with steel tyres having a high load concentration. Now that most vehicles have pneumatic tyres, these properties are less important in Britain.

The binding properties of the stone dust as measured by the cementation value have little relation to the other properties. A high cementation value was considered an advantage, but of secondary importance to crushing strength.

3.1.1.2 Modern unsurfaced roads

A study has been made of test procedures for weak rocks (Shergold and Hosking, 1963). Some shales and other weak rocks, when used in sub-bases and unsurfaced roads, break down excessively under compaction and the compacted material may then have reduced stability when wet. The investigation had the object of finding a test procedure to distinguish these rocks from others that did not break down excessively. The results showed that rocks unacceptable to engineers because of their breakdown under compaction could be rejected by imposing a maximum value of 40 for the aggregate impact value or a minimum value of 5 tonnes for the 10% fines value.

3.1.1.3 Roadbase and sub-base materials

a.   General

The main purpose of the lower layers of a modern road, the roadbase and the underlying sub-base, is to distribute the traffic stresses transmitted through the upper layers over a sufficient area of the underlying formation (subgrade) to avoid deformation. Although bound materials are better, unbound aggregates are commonly used for these layers because of their lower cost.

The requirements for these unbound aggregates are as much to do with transport, laying and subsequent trafficking by construction vehicles as with their function when the road is in service. Grading and plasticity are important; they must have adequate strength and resistance to attrition and, in addition, be resistant to frost action and other forms of weathering.

A number of tests have been specially developed in order to assess the durability of roadbase and sub-base aggregates. These include the slake durability test, the Texas ball mill test and the Washington degradation test. But in Great Britain the 10% fines test is preferred for this purpose.

The frost resistance of sub-base materials has been the subject of much research (e.g. Croney and Jacobs, 1967). This work was aimed at identifying frost-susceptible aggregates which could otherwise be used in an unbound form in roadbases and sub-bases at depths likely to experience frost penetration during the life of the road. It led to the specification of the frost heave test (Roe and Webster, 1984) and its subsequent adoption by the British Standards Institution in BS 812:Part 124 "Method for the determination of frost-heave".

Recent research (Webster and West, 1989) has shown that small proportions of additives can reduce the frost-heave of otherwise frost-susceptible gravels. Two per cent of Portland cement or bentonite was found to reduce frost-heave to almost zero, and the same percentage of hydrated lime reduced the heave by about one half. Strength was not reduced by these additions.

In Great Britain the main requirements used for these materials are those specified in the Department of Transport's "Specification for Highway Works" (Department of Transport, 1986). This specifies a range of permitted sub-base materials that includes two types of granular sub-base material (Type 1 and Type 2), wet-mix macadam and dry-bound macadam. Roadbase requirements are more demanding; in addition to bound materials, the use of wet-mix macadam and dry-bound macadam is permitted. The requirements are as follows:

i   Type 1 and Type 2 granular sub-base materials.

Type 1 material applies to crushed rock and similar crushed materials, whereas Type 2, used for the less heavily trafficked roads, covers a wider range of materials including sands and gravels. The percentage of material passing a 75 µm test sieve must not be more than 10 and the material passing a 425 µm test sieve must be non-plastic (this requirement is relaxed to allow materials with a plasticity index of less than 6 in the case of Type 2 materials). The strength is specified in terms of 10% fines value, which must not be less than 50 kN on saturated and surface dry aggregate. Additionally any material used within 450 mm of the road surface must be non-frost susceptible and any aggregate placed within

500 mm of concrete structures or cement-bound materials must have a soluble sulphate content not greater than 2.5 g/l.

ii Wet-mix macadam.

This must comply with BS 882:1983 "Aggregates from natural sources for concrete" and, additionally, have no more than 8 percent material passing a 75 µm test sieve. It should also comply with the same frost and soluble sulphate requirements as the granular sub-base materials.

iii Dry-bound macadam.

This must comply with the same requirements as wet-mix macadam except that there should be no material passing a 75 µm test sieve.

(*Author's note*: DTp's "Specification for highway works" is revised every few years, the information given in this book is from the 6th Edition.)

b.   *Research on compaction*

The Sand and Gravel Association's (SAGA) Research Team, in their collaborative programme with the Department of Transport, carried out a detailed investigation of the mechanical properties of graded aggregates (Pike, 1972; Pike and Acott, 1975). The main object of this work was to set standards for shear testing, but it also yielded a considerable amount of other useful information. It was also shown that aggregate type, particle-size distribution and moisture content all have an important effect on compactability. This work also included a study of the British Standard vibrating hammer compaction test (BS 1377).

Sieving tests were performed to assess the degree of degradation that occurred during compaction. It was found that both the resistance to crushing of individual particles and the grading strongly affected the degree of degradation.

The existing method of test was reviewed and new improved methods of assessing degradation from the results of sieving tests were developed. This work led to proposals for a modified vibrating hammer compaction test (MVHT). It also allowed the classification of the compactability properties of graded aggregates, properties that also influence the behaviour of cement- and bitumen- bound materials made from such aggregates.

c.   *Research on shear strength*

The SAGA DTp Research Team's work was primarily concerned with a study of methods of test for the shear strength of aggregates. A preliminary investigation into the use of the existing standard shear-box test for measuring the resistance to shear of a compacted mass of clean, graded aggregate (as used in Type 1 sub-bases) showed that this test was suitable for differentiating between various types of aggregate over a wide range of shear properties, and that the test placed the aggregates in an apparently logical order of performance. This investigation led to the design of a modified shear-box that allowed tests to be carried out at stresses up to 1 MN/sq.m. rather than the 160 kN/sq.m. that could be achieved with the existing equipment.

The first phase of an investigation using the shear-box to study the effects of aggregate type and grading upon the mechanical properties of graded aggregates showed that, at low levels of normal stress, measured shear strengths were sensitive to aggregate type but, at higher levels of normal stress (1 MN/sq.m.), particle angularity and roughness have little effect on shear strength. However grading, moisture content and density of packing were all found to have important effects on shear strength at all level of normal stress. The effects of these variables were quantified and related to the compactability of the experimental materials.

Further work (Pike 1973) showed that although the relationship between peak shear-stress and normal stress was non-linear, an arbitrary index could be formulated to distinguish between aggregates of different types. Later work led to the shear-box test procedure being simplified and standardised, and the precision determined.

It was envisaged that this standardised test could be useful to suppliers of sub-base materials and to contractors, for example, in evaluating the "trade-off" between cheaper, less stable, local materials with more expensive sub-bases available at some distance from the area in question. A threshold value of stability was identified, and this allowed broad classifications of basic petrological types of aggregates to be made.

d.   *Pilot scale compaction studies*

Granular sub-bases, which form the lowest layer of both flexible and rigid pavements, are under greatest risk of failure from construction traffic when the next layer of the pavement is being built. The usefulness of the shear-box test was therefore evaluated in a series of pilot-scale trials carried out by the SAGA team. Compacted aggregates were laid on a prepared clay sub-grade and trafficked by full-scale construction plant to test the validity of the classification obtained with the test.

During 1975 (Transport and Road Research Laboratory, 1975) results were obtained for a range of non-plastic (Type 1 sub-base) materials. As expected, flint sands and gravels were unable to support construction traffic, but limestone and gritstone gravels gave adequate performance. A flint gravel mixed with crushed limestone fines performed as well as a crushed granite and such mixtures may well show a cost advantage in the south east of England. Fig. 3 shows the relationship found between rut-depth and the number of passes of a vehicle on the six sub-base materials.

e.   *Research on the stability of non-plastic sub-bases*

A report (Pike *et al*, 1977) sets out methods of predicting the stability of non-plastic sub-bases under construction traffic. Simple tests on grading, angularity and petrological groupings of aggregates were found to be unreliable parameters of stability. Standard bearing tests such as the California bearing ratio (CBR) were shown to give results that are also unsatisfactory for this purpose.

Although it was found that the established Mohr-Coulomb shear-strength theory could not be applied, the modified 300 mm shear-box test was found to provide reliable classification of potential stability. This was demonstrated by comparing results of such tests on Type 1 sub-base materials with the performance of these materials in the pilot-scale trafficking trials described above.

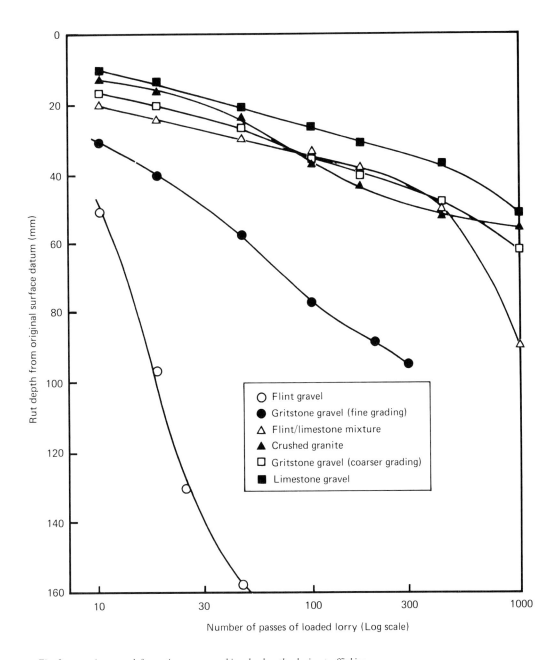

Fig. 3   Average deformations measured in wheel-paths during trafficking.

3.1.2 Concrete road construction

3.1.2.1 General

The main requirements of a road-building concrete are adequate strength, mechanical and chemical durability and, if the concrete is used as a road surfacing, adequate resistance to wear and polishing. In order to achieve these requirements the aggregates should have characteristics that will provide a mix that will harden to a suitable concrete. A knowledge of grading and water content is necessary in order to design the mix. Concretes are usually designed as a mixture of coarse and fine aggregates, but sometimes an "all-in" aggregate is used. Coarse aggregate consist of particles that are largely greater than 5 mm and fine aggregates have particles mostly smaller than this size.

Although there is little need for the strength of a coarse aggregate used in a concrete road to be greater than that of the mortar, reasonable strength is desirable. Chemical interaction with the binder is more likely than with bituminous binders and restrictions on levels of organic matter, certain iron and manganese compounds, gypsum and a number of other deleterious substances are therefore necessary.

Concrete aggregates should comply with the requirements for grading and shape that are laid down in the various specifications. The basic requirements are those specified in BS 882 for aggregates from natural sources and BS 1047 for blastfurnace slag aggregate. The fine aggregate in the mortar has been found to play a more prominent part than the coarse aggregate in governing the skid-resistance of a concrete surfacing and consequently there is less need to control the properties of the coarse aggregate than in bituminous road surfacings. Density and cleanliness affect the mixtures in the same way as in bituminous road materials.

3.1.2.2 Strength and mechanical durability

Research has shown (e.g. Clemmer, 1943; Graf, 1936 and Phemister *et al*, 1946) that, within wide limits, the strength of the aggregate had little effect on the crushing strength or flexural strength of the concrete made with it. It was considered to be important only with very weak aggregates or when a particularly strong concrete is required.

Resistance to abrasion was considered to be of more importance. For light traffic with pneumatic tyres it was found that adequate abrasion resistance could be achieved with aggregates with a crushing strength exceeding 12,000 lb/sq.in (83 MN/sq.m.) (aggregate crushing value not above 35). For heavy traffic with pneumatic tyres the recommended figures were 15,000 lb/sq.in. (103.5 MN/sq.m.) (aggregate crushing value not above 30) respectively. Where vehicles with iron tyres were employed it was suggested that the crushing strength should be at least 30,000 lb/sq.in (aggregate crushing value not above 17), and the abrasion value not to exceed 17 (the Dorry abrasion test). With such vehicles it was also found that brittle aggregates, e.g. some flints, were not satisfactory unless used in strong concrete.

The crushing strength and Dorry abrasion tests are not commonly used at the present time, but the aggregate crushing test remains. The aggregate impact test and 10% fines test have since been standardised and most present-day specifications for concrete aggregates specify aggregate impact values or 10% fines values. This is because they are more generally suited for testing the weaker materials that are adequate for concrete. A typical specification appearing in BS 882:1983 is reproduced as Table 4, the figures supersede the earlier maximum aggregate crushing values of 30 for concrete wearing surfaces and 45 for other concrete. The aggregate impact test is used as an alternative to the 10% fines test.

Wartime activities included studies of the choice of aggregates for roads carrying tracked

TABLE 4
Limiting values on mechanical properties for different types of concrete (BS 882:1983)

Type of concrete	Ten-per-cent fines value (kN) (not less than)	Aggregate impact value (%) (not exceeding)
Heavy duty concrete floor finishes	150	25
Pavement wearing surfaces	100	30
Others	50	45

41 N/mm^2

vehicles (Road Research Laboratory, 1949). This work indicated that "the type of aggregate has an important effect on the later stages of wear of medium and low strength concrete, best results being given by igneous rocks and worst results by flint gravels The type of aggregate has, however, little effect on the wear resistance of concrete having a strength of more than 6,000 lb/sq.in. (41 MN/sq.m.)" This may be of local interest in some areas but tracked vehicles are not permitted to use public highways. However there is a potential problem in the use of studded tyres. If they come into general use in Great Britain, it would lead to a re-evaluation of highway surfacing practice.

3.1.2.3 Chemical durability

a. Alkali-aggregate reactivity

Weathering has not generally been considered to be a problem with roadstones of adequate strength. However instances of chemical instability have been found to occur with certain mineralogical types of rock. Problems with chert were first encountered in America (e.g. Sweet and Woods, 1942) and other American workers (e.g. Hansen, 1944 and Stanton *et al*, 1942) made early studies of the alkali-aggregate reaction. In addition problems were encountered with certain dolerite rocks (Phemister *et al*, 1946).

More recent studies of these problems have identified three main types of alkali-aggregate reactivity. These are:

i   Alkali-silica reaction. Reaction between alkalies and some forms of silica, such as chalcedony, chert, natural glass and opal gives rise to the formation of an alkali-silicate gel. This absorbs moisture, swells and disrupts the concrete.

ii Alkali-silicate reaction. Expansive reactions can also occur between alkalies and some types of silicate materials such as chlorite, phyllite and vermiculite (Gillott *et al*,1973).

iii Alkali-carbonate reaction. Deleterious reactions have been observed between the alkalies in cement and some carbonate rocks (e.g. Poole and Satiripoulos, 1981). This problem appears to be associated mainly with dolomitic aggregates.

Petrographic examination can help to identify aggregates that might be prone to alkali-aggregate reaction, and a number of test methods have been formulated. The American Society for Testing and Materials (ASTM) have published a method for petrographic examination (ASTM C295), a chemical method (ASTM C289), a mortar-bar method (ASTM C227) and a rock cylinder method (ASTM C586). Recommendations for minimising the risk of alkali-silica reactivity have also been published (Hawkins *et al*, 1983).

b.   *Alkali-silica reaction in roads*

Alkali-silica reaction takes place in concrete when an alkali solution, originating either from pore solutions in the cement paste or from some other source, reacts with some types of silica in the aggregate to form an expansive gel which exerts a pressure that can crack both aggregate particles and cement paste.

West and Sibbick (1988a) have investigated the occurrence of alkali-silica reaction on concrete roads. They found evidence of this reaction in two of the fourteen concrete roads that were examined. One of these, on Motorway M40 in Oxfordshire, was then studied in detail (Sibbick and West, 1989). The main findings were:

i   Evidence of damage by alkali-silica reaction. This was closely associated with the joint edges of the slabs and was considered to have followed some other primary cause of damage.

ii   The reactive aggregate was a highly porous flint occurring in the Thames Valley river gravel used in the concrete.

iii   Alkali derived from road de-icing salt was considered to have added to the alkali content of the pore solution.

iv   Poorly sealed and non-functioning joints, together with the cracks associated with them, were important factors in allowing the ingress of water and salt.

v   Recommendations to minimise the reaction (other than the use of non-reactive aggregates) were adequate sealing of joints, the minimal use of de-icing salt and air entrainment.

Work on a further nineteen samples (West and Sibbick, 1989a) showed that five of the total of thirty three samples exhibited alkali-silica reaction. The reactive aggregates in four of these were flint gravels, in the fifth the reactive aggregate was a siltstone. This work has shown that whilst alkali-silica reaction is not a major problem in existing concrete roads, it can occur when a road becomes badly cracked and water gains access to a potentially reactive aggregate. Petrographic examination of thin sections of concrete may be carried out to identify alkali-silica reaction where it is suspected. The presence together of reactive aggregate, micro-cracks and alkali-silica gel is indicative of the reaction. These workers have concluded (West and Sibbick, 1989b) that, in order to reduce the likelihood of alkali-silica reaction occurring in concrete roads, it is important to use non-reactive aggregates and a low-alkali cement. It is also necessary to ensure that joints are soundly constructed, sealed and well-maintained. Additionally the use of de-icing salt should be kept to a minimum consistent with road safety. Air-entrained concrete was found to suffer less cracking than normal concrete because the presence of voids allows the gel to expand.

c.   Drying shrinkage

Problems with some dolerite rocks have been found to be attributable to a high drying shrinkage of the aggregate. Concrete tends to shrink on drying, but the resistance to shrinkage of most aggregates tends to limit the extent. Some aggregates, particularly those containing clay minerals, resulting from the weathering of constituent minerals such as olivine, exhibit a high drying shrinkage and concretes made with them may show excessive shrinkage. The problem is most serious in buildings, where more drying takes place and shrinkage can cause structural failure. Problems have also been encountered with road bridges. Less drying occurs in roads and shrinkage does not have such catastrophic results. However the consequent cracking can lead to greater risk of frost damage and the corrosion of steel reinforcement.

An aggregate shrinkage test has been developed in Britain that measures the shrinkage of a standard concrete bar made with an aggregate. This has been generally referred to as the "Digest 35 test". This digest (Building Research Station, 1963) lists four categories of aggregate shrinkage that are based on test results, it comments on the types of aggregate that qualify for each group and gives recommendations for their use. BS 812:Part 120:1989 specifies two methods of testing and classifying aggregate shrinkage; both are based on the Digest 35 method.

A methylene blue dye test has also been studied and has value as a "screening test" (Hosking and Pike, 1985).

d.   Staining

Another chemical problem can arise through the presence of impurities such as pyrite (a mineral consisting of iron sulphide, also known as iron pyrites); the oxidation of these impurities can lead to unsightly staining and surface blemishes.

e.   Retardation

Some impurities in concrete aggregates seriously retard the setting rate of concrete. These include some organic materials and lead and zinc oxides. The latter are not common contaminants of aggregates, but problems with organic matter have been more common. A simple colorimetric test was included in BS 812 for many years (the organic impurities test), but was superseded by a pH test in 1967 because the colorimetric method did not correctly classify all organic contaminants. However the pH test was omitted from the later revision of BS 812 because it, too, was found to be unsatisfactory. A version of the colorimetric test is used in the United States of America (ASTM C40) and other parts of the world.

f.   Sulphates

The disruption to concrete caused by aggregates containing an excessive amount of sulphates is uncommon in Great Britain, but has given trouble overseas particularly in the Middle East. Magnesium and sodium sulphates are the usual cause of the trouble. Calcium sulphate is less soluble and less likely to react with the cement compounds, however in the form of gypsum, it can give rise to problems because of its "unsound" nature. BS 812:Part 118:1988 specifies a procedure for determining the sulphate content of aggregates.

TABLE 5
Maximum chloride content (BS 882:1983)

Type or use of concrete	Maximum total chloride content expressed as a percentage of chloride ion by mass of combined aggregate
Pre-stressed concrete Steam-cured structural concrete	0.02
Concrete made with cement complying with BS 4027 or BS 4248	0.04
Concrete containing embedded metal and made with cement complying with BS 12	0.06 for 95 per cent of test results, with no result greater than 0.08

g. *Chlorides*

Sodium chloride (common salt) and other chlorides are sometimes found in concrete aggregates, particularly those of marine origin. Their presence can cause problems with reinforced concrete by attacking the steel, both reducing its effectiveness and causing disruption by the expansion due to corrosion. The damage caused to steel bridge members is a particularly serious consequence of excessive chlorides. BS 812:Part 117:1988 specifies a method of determining the water-soluble chloride salt content of aggregates. It should be noted that damaging chlorides can come from sources other than the aggregate, examples are de-icing salt, exposure to sea water and accelerators in the cement.

Appendix "C" of BS 882:1983 provides a table of suitable chloride limits (this is reproduced as Table 5) for various applications. The values should not be exceeded unless calculations made on measured values show that the total chloride of the concrete meets the specification for the concrete.

h. *Shell content*

The presence of shells and/or shell fragments in marine and coastal gravel deposits can be detrimental to concrete affecting workability and entrapping air. BS 812:Part 106:1985 specifies a method for determining the shell content of a coarse aggregate. BS 882:1983 includes limits for the shell content of aggregates; these are reproduced in Table 6.

TABLE 6
Limits on shell content (BS 882:1983)

Size	Limits on shell content (%)
Fractions of 10 mm single size, or of graded or all-in aggregate that are finer than 10 mm and coarser than 5 mm	20
Fractions of single sizes of graded or all-in aggregate that are coarser than 10 mm	8
Aggregates finer than 5 mm	no requirement

i.   Special requirements for slag aggregate

Because some blastfurnace slags can be of low density, a lower limit of 1250 kg/cu.m. is specified in the British Standard for air-cooled blastfurnace slag aggregate (BS 1047). Poor stability can also be a problem and BS 1047 requires that the aggregate should comply with certain stability requirements namely, absence of sulphur unsoundness, iron unsoundness and lime unsoundness (also termed "falling" or "dusting"). Test methods for these properties are given in Appendix A of BS 1047:1973.

BS 1047 specifies limits for sulphur (total sulphur not to exceed 2.0%) and sulphate (total sulphate ion not to exceed 0.7%). Additionally water absorption is limited to a maximum of 10%.

Although drawn up as a standard for concrete aggregates, BS 1047 is also used for other purposes. For example, it is widely used for specifying blastfurnace slag for bituminous materials.

3.1.2.4 Skid-resistance

A detailed account of studies of the part played by aggregates in imparting skid resistance to concrete road surfacings is given in Chapter 5.

3.1.2.5 Physical properties

Knowledge of a number of physical properties is necessary in order to design a concrete mix. These are particle density, bulk density, water absorption, moisture content and grading. As well as being useful in mix design, knowledge of the particle density and water absorption can indicate the suitability of the aggregate for concrete construction. The physical properties of particle shape, surface texture and the presence of clay, silt and dust also influence the quality of the concrete. These properties are discussed in the following paragraphs:

a.   Particle density and water absorption

Particle density, formerly known as specific gravity or relative density, and water absorption of aggregates are measured at the same time by one of the methods specified in BS 812:1975. The Standard for blastfurnace slag aggregate (BS 1047) requires the water absorption not to exceed 10%.

b.   Bulk density

Bulk density is measured by the method specified in BS 812:1975. It is commonly used to enable concrete mixes that have been specified by volume to be batched by weight. It also gives an indication of the void content of an aggregate. Knowledge of bulk density is of particular value when using slag aggregates; steel slags can have high densities whereas those of some blastfurnace slags are low. BS 1047 requires that the bulk density of blastfurnace slag aggregate should not be less than 1250 kg./cu.m.

c. *Moisture content*

Knowledge of the moisture content of both the coarse and fine aggregates to be used is necessary in order to adjust the water content of a mix correctly. A number of suitable test methods appear in BS 882. The methods usually employed are by use of the siphon can, steelyard moisture meter or buoyancy moisture meter.

d. *Grading*

The gradings of both the coarse and fine aggregates need to be known in order to achieve the best quality mix. BS 812:Part 103:1985 specifies methods of determining the particle size distribution of aggregates. BS 882 sets out grading limits for both coarse and fine aggregates. Coarse aggregates consist of material mainly retained on a 5 mm test sieve, whereas fine aggregates are those mainly passing this sieve size. (This contrasts with the 3.35 mm test sieve used for bituminous materials). BS 882 coarse aggregates can be "single-sized", "graded" or "all in". Fine aggregates were formerly specified according to four rather rigid grading zones, but this system was revised in BS 882:1983 so as to provide a more flexible arrangement. Three categories, coarse, medium and fine ("C", "M", and "F"), replaced the four with a measure of overlap. This classification allows better use of our national resources of fine aggregates. In addition to these three categories, the specification also allows the use of other gradings provided that "the supplier can satisfy the purchaser that such materials can produce concrete of the required quality". The three BS 882:1983 categories are reproduced in Table 7.

e. *Particle shape*

Particle shape is important in that excessive amounts of flaky or elongated material in aggregates can affect the workability of the concrete. BS 812:Part 105 gives methods of determining the flakiness index and elongation index of an aggregate. Limits for flakiness index are given in BS 882:1983 for coarse aggregate. These are that "the flakiness index of the combined coarse aggregate shall not exceed 50 for uncrushed gravel and 40 for crushed rock or crushed gravel". A note states that "for special circumstances, e.g. for pavement wearing surfaces, a lower flakiness index may be specified".

TABLE 7
Fine aggregate grading limits (BS 882:1983)

	Percentage by mass passing BS sieve			
	Overall	*Additional limits for grading*		
	limits	*C*	*M*	*F*
Sieve size		*(Coarse)*	*(Medium)*	*(Fine)*
10.00 mm	100	-	-	-
5.00 mm	89-100	-	-	-
2.36 mm	60-100	60-100	65-100	80-100
1.18 mm	30-100	30-90	45-100	70-100
600 µm	15-100	15-54	25-80	55-100
300 µm	5-70	5-40	5-48	5-70
150 µm	0-15*	-	-	-

* Increased to 20% for crushed rock fines, except when they are used for heavy duty floors.

f.   Angularity, roundness and compactability

At one time BS 812 included a determination of angularity number. The test has since fallen into disuse. It measured the angularity of an aggregate (or conversely its roundness) by assessing the amount of voids remaining in a sample of aggregate after it had been subjected to a rigorous compacting procedure in a metal vessel. The test evaluated compactability rather than angularity because the surface roughness of the aggregate also affected the compaction.

g.   Smoothness, roundness and the flexural strength of concrete

Research (Director of Road Research, 1956) on the failure of concrete test specimens showed that the flexural strength was not uniquely related to the compressive strength for all of the concretes, but a good general relation was found between the flexural strength and the stress at which cracks were first detected during the compression tests. Both these properties were found to be dependent on the type of aggregate; in general, the smooth rounded stones produced concrete of lower flexural strength, which cracked at a lower compressive stress than that produced by a more angular and rough aggregate.

h.   Clay, silt and dust

The amount of fine material finer than 75 μm in an aggregate can have a considerable influence on the properties of a concrete. The effect becomes greater as the particle size diminishes and so clay can be far more harmful than silt. BS 812 defines "clay, silt and dust" collectively as material passing a 75 μm test sieve, and BS 812:Part 103 includes methods of determining the amount present in a sample. Some fine aggregates, such as crushed rock fines, contain coarser fine material than others, so different limits are necessary. BS 882:1983 lists appropriate limits for the decantation test, these are reproduced in Table 8.

BS 882:1983 allows the field setting test to be used as a screening test. If the amount of clay and silt as determined by this test is greater than 10% by volume, a decantation test is required unless a figure greater than 10% by volume represents not more than 3% by mass.

TABLE 8
Limits for clay, silt and dust (BS 882:1983)

Aggregate type	Quantity of clay, silt and dust (maximum percentage by mass)
Uncrushed, partially crushed or crushed gravel	1
Crushed rock	3
Uncrushed or partially crushed sand or crushed gravel fines	3
Crushed rock fines	15*
Gravel all-in aggregate	2
Crushed rock all-in aggregate	10

* 8 for use in heavy duty floor finishes

The presence of clay, either in large particles or as an adhering coating can be detrimental to the concrete. At present there is no British Standard test for this property, but the American Standard ASTM C142 gives a test method for clay lumps and friable particles.

i. *Compactability and concrete mix design*

The compactability of the graded aggregate mixture used in concrete has an important influence on the properties of the concrete (Pike, 1972; Pike and Acott, 1975). The modified vibrating hammer test (MVHT), developed to measure compactability, was used in a study of the mix-design of concretes. Both the results of the MVHT and the workability of the concrete were found to be influenced by the particle geometry (angularity, shape, texture, etc) of the aggregate. MVHT values are useful in forecasting the water demands of concretes for given levels of workability when using different types and grading of aggregate.

j. *The effect of the properties of coarse aggregate on the workability of concrete*

A study (Kaplan, 1958) has been made of the effects of shape, surface texture and water absorption on the workability of concrete. Workability was assessed in terms of compacting factor (BS 1881:1952). It was found that:

i Changes in the angularity of coarse aggregates had a greater effect on the workability than changes in flakiness index.

ii Increased angularity number and/or flakiness index led to a reduction in workability (see Figs. 4 and 5).

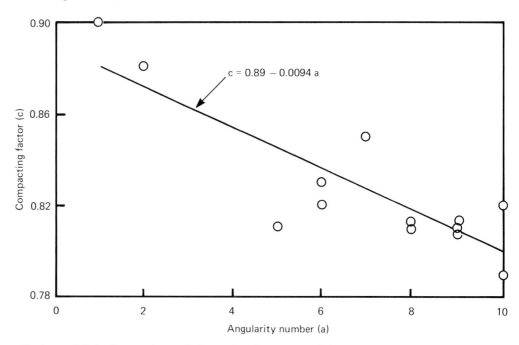

Fig. 4 *Relation between the angularity number of aggregate and the compacting factor of concrete made from it.*

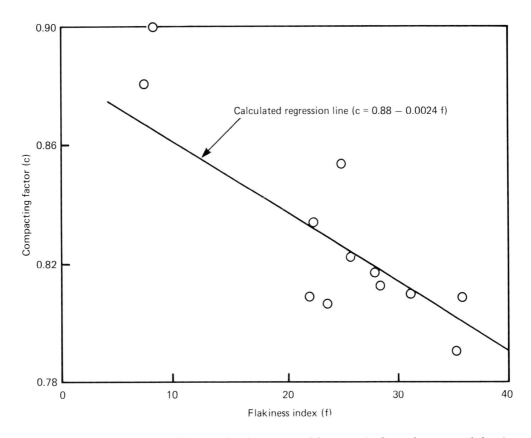

Fig. 5   Relation between the flakiness index of aggregate and the compacting factor of concrete made from it.

iii   No correlation was found between surface texture measurements (Wright, P J F, 1955) and workability over a wide range of textures.

iv   No correlation was found between water absorption over the range 0.5% to 1.4% and the workability.

3.1.2.6 Department of Transport's requirements

The Department of Transport's "Specification for highway works (6th edition)" (Department of Transport, 1986) requires that aggregates for all pavement concrete shall be either natural material complying with BS 882 or crushed air-cooled blastfurnace slag complying with BS 1047. If pulverised fuel ash (PFA) is used, it must comply with BS 3892. Other requirements are:

i   The nominal size must not exceed 40 mm. This is reduced to 20 mm if spacing of longitudinal reinforcement is less than 90 mm.

ii   Chloride ion content must satisfy Appendix "C" of BS 882 (i.e. Table 5).

iii No more than 25% by mass of acid soluble material (BS 812) in either fraction greater or less than 600 µm shall be used in the top 50 mm of surface slabs.

iv Limestone rock aggregate used in the top 50 mm must give a value of not less than 53 in the accelerated wear test.

v Requirements are given to control alkali-silica reaction.

3.1.3 Bituminous road construction

3.1.3.1 Coated macadam and asphalt

The general requirements of aggregates for bituminous road construction are as follows:

a. Type of aggregate

In the past, specifications tended to differentiate more than at present between different types of aggregate and binder. BS 4987, for example, replaces several earlier British Standards for coated macadams. One set dealt with tar-bound and another with bitumen-bound materials, and there were different standards for crushed rock, slag and gravel. The latest versions of the British Standards for coated macadam and hot rolled asphalt (BS 4987:1988 and BS 594:1985 respectively) each divide aggregates into the same four types, and give a few special requirements and notes appropriate to each. These requirements are dealt with more fully in later paragraphs. The four aggregate classifications are:

i Crushed rock of one or more of the following groups: basalt, gabbro, granite, gritstone, hornfels, limestone, porphyry or quartzite. (It is appreciated that certain satisfactory aggregates may not belong to any of these groups, and can be used with the approval of the purchaser).

ii Gravel of one or more of the groups in (i) or flint, crushed or uncrushed, or combinations of both. With the exception of limestone gravels, it is required that 2% by mass of total aggregate shall be either hydrated lime or Portland cement as filler. (This is to reduce problems of adhesion between the aggregate and the bituminous binder - see below).

iii Blastfurnace slag. This is required to comply with BS 1047 "Air-cooled blastfurnace slag coarse aggregate for concrete", because of the particular problems that can be associated with blastfurnace slags.

iv Steel slag, either electric arc furnace slag or basic oxygen slag. This is required to be weathered "until it is no longer susceptible to falling" and the compacted bulk density shall be between 1700 kg/cu.m. and 1900 kg/cu.m.

b. Mechanical properties

The aggregate to be used in a flexible road mixture should have sufficient strength to be able to withstand the stresses imparted by traffic and construction machinery. Because the load is spread over a greater area as the distance increases from the surface, the strength requirement is different for each layer. The British Standard (BS 812) aggregate impact and 10% fines tests are more commonly used for this purpose in Great Britain. Adequate aggregate abrasion value almost invariably implies adequate strength. The American Los Angeles test (ASTM C131 and ASTM C535) is also used world-wide for this purpose. Neither BS 4987 for coated macadam nor BS 594 for hot-rolled asphalt give limits for strength, but do refer to the 10% fines test for this purpose.

c.   Soundness

A road aggregate should be able to resist the effects of the climate in which it is to be used. Although the binder coating tends to inhibit breakdown through weathering, aggregates, particularly those to be used in the surfacing or in uncoated underlying layers, should be able to withstand the physical and chemical actions of weathering. Frost resistance is also important in countries such as Great Britain.

A study (Hosking and Tubey, 1969) was made of the problems that arise in regions deficient in good-quality aggregates. It included the assessment of low-grade but sound aggregate, and of apparently sound aggregates which rapidly disintegrate in service. It was concluded that the modified aggregate impact test and modified 10% fines tests on water-saturated samples could be used to assess sound aggregates, but that no really satisfactory test had been devised that will detect all types of unsound aggregates.

In a number of full-scale road surfacing experiments carried out by the Road Research Laboratory, notably those at Blackbushe (Hosking, 1967b) and Derby Ring Road (Brown, 1967), a disappointing performance was given by aggregates that proved to be unsound.

Tests that have been found to be of value in assessing the soundness of aggregates include the sodium sulphate soundness test and the similar magnesium sulphate soundness test. Standard test procedures for both are given in the American standard ASTM C88 and the magnesium sulphate version in BS 812:Part 121, the 10% fines durability factor test is also specified in this British Standard. Another test that has been used to assess soundness is the methylene blue dye test. More information on these tests is given in Chapter 4.

d.   Grading and shape

Aggregates should comply with the optimum grading limits that have been established for flexible road mixtures. These are laid down in the various appropriate specifications such as BS 594 and BS 4987. The shape and texture of aggregates can also affect the strength, workability and resistance to deformation of these mixed materials. Controlled grading and shape are also required of aggregates for use as chippings for surface treatment (either surface dressings or chippings in asphalt) in order to give a suitable surface texture and to facilitate laying (e.g. BS 63). The methods of test generally used in Great Britain are sieve analysis (BS 812:Part 103), the determination of fine materials (BS 812:Part 103.2) and the determination of flakiness index (BS 812:Part 105.1).

Coarse aggregates for bituminous mixtures are defined as those substantially retained on a 3.35 mm test sieve, and fine aggregates as those substantially passing this sieve. This contrasts with the 5 mm test sieve size used for the same purpose with concrete aggregates.

The same flakiness index limits are specified in BS 594 and BS 4987; these are 45% for crushed rocks and 50% for gravels. BS 594 allows up to 5% material finer than the 75 μm test sieve for crushed rocks, but a lower limit of 3.5% is imposed for gravels. BS 4987 specifies a limit of 1% for gravels. Both sedimentation and decantation tests are allowed but, in the event of a dispute, the sedimentation test is to be employed.

e.   Resistance to abrasion and polishing

Although of no great importance in underlying layers, good resistance to abrasion and polishing are necessary for the aggregates that are exposed in the surface. These properties are necessary to ensure adequate durability, texture depth and skid resistance. The degrees of abrasion and polishing resistance that are required are governed by the type of road site and the amount of traffic. Adequate abrasion resistance is also necessary to avoid breakdown during mixing. The test methods used in Great Britain are the British Standard polished-stone value test and the aggregate abrasion test, both of BS 812. More detailed information is given in Chapters 5 and 6 (See also 3.1.3.2 below).

f.   Density

Because flexible road mixtures are usually batched by weight, the density of the aggregate used needs to be taken into account when formulating mixtures. Knowledge of the bulk density (or of particle density) of an aggregate is therefore required, particularly for the lighter-weight blastfurnace slags which require appreciably more binder than an average aggregate. Knowledge of density is also required to assess the rate of spread of chippings for surface treatment.

Lower density/higher porosity implies lower strength and greater frost susceptibility. However some low density/high porosity aggregates give a satisfactory performance and the type of aggregate should not be excluded by these tests alone.

g.   Cleanliness

Aggregates for bituminous materials should be reasonably free from deleterious matter such as clay, organic matter and excessive dust. An excess of these materials can seriously impair the performance of the mixture. Both BS 594 and BS 4987 point out that there is no acceptable British Standard method of determining clay content, but that the sedimentation test of BS 812 does give some indication of clay content.

h.   Adhesion to binders

Provided the aggregate is dry and not unduly dusty, all aggregates can be coated with road tar or bitumen by the usual mixing process and good adhesion is obtained. Water, however, has a greater affinity to stone surfaces than has either tar or bitumen, and when a coated stone comes in contact with water there is a tendency for the binder to be displaced from the surface of the stone by the water. The degree to which the binder will adhere to the stone in the presence of water is determined partly by the properties of the binder and partly by the properties of the stone. The properties of the stone which are important in this connection are the surface texture and the mineralogical composition. In general, aggregates of acid rocks (those containing a predominance of acid minerals such as quartz) adhere less well than those of basic rocks such as limestones (Mathews, 1958).

The most satisfactory way of assessing the adhesive properties has been by comparison of the behaviour of aggregates when subjected to immersion wheel-tracking tests (Mathews and Colwill, 1962). The addition of certain wetting agents to the binder or the addition of hydrated lime or Portland cement to the fine aggregate had been found to reduce stripping. Both BS 594 and BS 4987 require that 2% of the total aggregate content of non-limestone gravels shall consist of hydrated lime or Portland cement as filler. They also include a note saying that the risk of stripping may also be reduced by the addition of adhesion agents.

i.   Appearance

Aggregates are sometimes required to provide a particular appearance to a road surface. White aggregates are required as an aid to road safety or to improve lighting conditions, such as in tunnels. White aggregates were at one time in great demand for road markings but have been largely replaced by ballotini (small glass beads) which impart better visibility but are less skid-resistant.

Coloured aggregates are also required for particular purposes. Red is commonly used for aesthetic purposes and different colours are found to be of value in delineating particular routes or areas such as bus stops.

Calcined flint has been the most commonly used white aggregate in Great Britain and is usually sold under trade names. Other aggregates that have been available are synthetic materials such as "Synopal" and white anorthosites or labradorites (rocks with a high proportion of white feldspar). No strongly coloured natural aggregates are available, the best being those of reddish hue. Road Note 25 (Road Research Laboratory, 1966) gave information on the white and coloured aggregates available at the time of publication. A method of measuring the colour of aggregates has also been published (Hosking and Ritson, 1968).

j.   Clean, hard and durable

The requirement "clean, hard and durable" has been included in standard specifications since early days and still remains in some specifications such as the Department of Transport's Specification for Highway Works (Department of Transport, 1986) for bituminous materials. Hitherto, definition of these terms has been left to the discretion of the engineer responsible for specifying, ordering and accepting these materials.

Recent research has been carried out (Bullas and West, 1991) with the object of defining these terms and determining quantitative criteria for each of these properties for aggregates used in bitumen macadam roadbase. This work has led to a cleanliness test based on the determination of material passing the 63 µm sieve of BS 812:Part 103:1985. The proposed limit of acceptability is 1 per cent. Recommended means of determining whether an aggregate is "hard" are the 10% fines and the aggregate impact tests; the recommended limits are 140kN or more and 23 or less, respectively. The magnesium sulphate soundness test is recommended for determining whether an aggregate is "durable": the recommended acceptability limit for this test being a value of 75 or more.

3.1.3.2 Bituminous mixture experiments

The following full-scale experiments were amongst the earliest carried out by the Road Research Laboratory that included a study of the behaviour of aggregates. Later experiments included studies of resistance to skidding and are described in Chapter 5.

a.   Local aggregates experiments

The requirements of aggregates for use in bituminous and tar carpets were studied by a series of full-scale road experiments carried out by the Road Research Laboratory during the 1940s and 1950s (e.g. Director of Road Research, 1952). The object was to find the best compositions for carpets made with locally available aggregates under local traffic and climatic conditions. Examples were the use of limestones in Somerset, Derbyshire and Monmouthshire; slag in Huntingdonshire; gravel in Staffordshire and Middlesex; and basalt, slag and granite in Ayrshire. Between 50 and 200 sections of different compositions were laid at each site. These experiments demonstrated the need to vary the mix to suit local aggregate traffic and climate, and paved the way to present-day specifications for bituminous road surfacing materials.

b.   Fine aggregates in dense bituminous surfacings

The full-scale road experiments on the Great North Road, A.1, and the Yaxley-Farcet road, B1091, in Huntingdonshire were begun in 1954 to test the laboratory findings on the effect of changing the type of fines in dense bituminous surfacing mixtures. These roads carried about 35,000 and 4,000 tonnes of traffic per day respectively in 1955. It was concluded (Director of Road Research, 1959) that both on heavily- and on lightly-trafficked roads, satisfactory results were obtained over a period of five years with rolled asphalts made with either crushed-rock or slag fine aggregate.

For the dense tar surfacings it was found that good results were obtained with blastfurnace slag aggregate over a range of binder contents considerably wider than can be used with sand as the fine aggregate. Also that with crushed-rock fine aggregate the surfacings were in a more satisfactory condition over a wider range of binder contents than were the mixtures made with sand. However the surface texture was rather variable.

3.1.3.3   Chippings for surface dressings and hot-rolled asphalt

a.   Early recommendations for aggregates in bituminous surfacing materials

Recommendations for aggregates for bituminous materials were proposed in the Road Research Laboratory's Special Report No.3 (Phemister *et al*, 1946). They were was based on a review of research (especially that in the USA (Clemmer, 1943)) and experience in Great Britain. It was recommended that reasonable practice would be to use aggregate with a crushing strength exceeding 15,000 lb/sq.in (103 MN/sq.m.)(or an aggregate crushing value of 30 or less) in the body of the surfacing and exceeding 25,000 lb/sq.in. (172 MN/sq.m.) (or an aggregate crushing value of 20 or less) in the wearing course. For bituminous surfacings on concrete, higher quality aggregate was recommended.

Adequate resistance to impact, abrasion and other traffic stresses were also regarded as essential.

However it was thought that a stone with an adequate strength would be unlikely to fail by impact. Good abrasion resistance was considered to be particularly necessary under intense traffic or where there were track-laying vehicles.

b.   *Crushing under the roller*

Fig. 6 shows the correlation that was obtained in a comparison of the Los Angeles abrasion values of a number of aggregates and the crushing that occurred under a 10 tonne roller expressed as the percentage of material produced that passed a No. 10 (about 1.5 mm) test sieve (Shelburne, 1940).

A laboratory investigation in Great Britain was also carried out on dry stone in order to obtain quantitative data on the relative importance of the various factors affecting crushing of aggregates under the roller (Director of Road Research, 1953). The results of this investigation showed that the degree of crushing occurring depended largely on the strength of the aggregate (as measured by the aggregate crushing test) and on the weight and number of passes of the roller. The grading and shape of the aggregate had less marked effects; about 20 per cent more crushing occurred in 1/2 inch (12.5 mm) single-sized aggregates of a poor shape (flakiness index = 40) and containing 15 per cent oversize and 30 per cent undersize, than occurred in those of good shape (flakiness index = 15) and with no oversize or undersize.

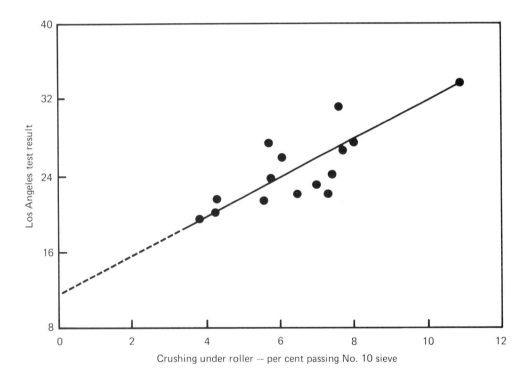

Fig. 6   *Correlation of Los Angeles test with crushing under 10 tonne roller.*

The strength needed to withstand rolling by a roller with steel wheels is greater than is needed for subsequent trafficking by vehicles with pneumatic tyres. This means that the introduction of rollers with rubber tyres has allowed the use of weaker aggregates, such as the highly polish-resistant gritstones, in road surfacings, so contributing to greater road safety.

c.   The Haddenham surface dressing experiment

This trial comprising 48 sections, each 100 m long, of surface dressing laid with the co-operation of the Buckinghamshire County Surveyor on the Class III road linking Thame with Haddenham. Six typical road aggregates were selected for test, together with two types of stone which were softer than would normally be used. Each aggregate was laid with three rates of spread of tar, the experiment being planned to give duplicate sections for each material and for each rate of spread.

Observations made during the gritting of the surface showed that the amount of crushing under a 10 tonne roller depended on the strength of the aggregate as measured by the aggregate crushing test and upon the extent to which the aggregate deviated from being a single-sized material.

Tests made when these surface dressings were laid showed that although the presence of the binder considerably reduced the amount of crushing that occurred under the roller, the relative amount of crushing that occurred with different stones was about the same with and without the binder.

After two years of trafficking (Director of Road Research, 1953) the results showed that although the different types of stone behaved differently, some breaking down more than others, all the surfacings were still in good condition. The degree of crushing occurring under traffic was found to correlate with the results of the aggregate crushing test except in the case of the slag.

None of the eight stones had polished, and it became evident that under such light traffic the weaker of the aggregates used, which would have hitherto been rejected by many engineers, would provide many years of useful life.

d.   Experiments on the effect of grading of chippings

Three collaborative full-scale road experiments by the Road Research Laboratory, the British Granite and Whinstone Federation and the Road Surface Dressing Association, were started in 1969 (Road Research Laboratory, 1969) to ascertain whether the tolerances for single-sized chippings given in BS 63:1951 were still valid for chippings for high-speed roads. The experiments were on the A46 at Shilton, Warwickshire, the A31 at Cadnam, Hampshire, and the A4 at Littlewick Green, Buckinghamshire.

These experiments showed that good quality surface dressings could be obtained with chippings of "single-sizedness" conforming with BS 63. There was also some evidence that the more the chippings approached a true single-size, the better the texture depth. It seems likely that the widely held belief that more single-sized chippings give a better surface dressing was probably because of the influence of factors like cleanliness rather than because of the "single-sizedness" itself.

e.   Wind-screen damage

The problem of wind-screen damage from loose chippings on surface dressings (Road Research Laboratory, 1971) has been shown to be most acute immediately after the work is done. Abnormally high rates of breakage were generally associated with the incorrect specification of chipping size for the traffic conditions. Road Note 39 (Road Research Laboratory, 1965) was published to cater for the need for different specifications to be prepared for each lane of multi-lane roads in order to minimise loss of surface texture by embedment of chippings in the road by heavy vehicles. The adoption of standardised methods of aftercare including vigorous sweeping assist substantially in reducing windscreen breakage.

f.   Present day specifications

The properties of aggregates used as chippings for surface dressing or in hot-rolled asphalt play a dominant part in determining the skid resistance of the surfacing. The requirements with respect to polishing resistance and wear resistance are considered later in Chapters 5 and 6. Apart from these, requirements for single-sized aggregates for surface dressing are specified in BS 63:1987. This specification replaces the earlier separate requirements for crushed rock and gravel respectively published in BS 63:1967 and BS 1984:1967. BS 63:1987 sets out two sets of requirements for chippings, one for general use and the other, less stringent, for lightly trafficked roads. The grading requirements are reproduced in Tables 9 and 10. In addition to these grading requirements there is also a maximum flakiness index limit of 25 and a minimum percentage of specified size of 65 per cent for the surface dressing chippings. For lighter traffic, the flakiness index maximum is relaxed to 35 and, for 10 mm and 6 mm chippings, the minimum percentage is reduced to 60. A strength requirement of 160 kN for the 10% fines test is specified for chippings for surface dressing, but is relaxed to 100 kN for blastfurnace slags.

Requirements for chippings for use in hot-rolled asphalt are more strict than those for surface dressing. These are specified in BS 594:1985 where the pre-coated chippings must comply with the grading of Table 11, must be of either 20 mm or 14 mm nominal size and have a flakiness index not exceeding 25. Additionally, where minimum texture depth requirements are needed

TABLE 9
Requirements for grading for surface dressing aggregates (BS 63:1987)

| BS test sieve | *Nominal size of aggregates* | | | |
| | 20 mm | 14 mm | 10 mm | 6 mm |
	Percentage by mass passing BS test sieve			
28 mm	100	-	-	-
20 mm	85-100	100	-	-
14 mm	0-35	85-100	100	-
10 mm	0-7	0-35	85-100	100
6.3 mm	-	0-7	0-35	85-100
5 mm	-	-	0-10	-
3.35 mm	-	-	-	0-35
2.36 mm	0-2	0-2	0-2	0-10
600 µm	-	-	-	0-2
75 µm	0-1	0-1	0-1	0-1

TABLE 10
Requirements for grading for surface dressing aggregates for lightly trafficked roads (BS 63:1987)

BS test sieve	Percentage by mass passing BS test sieve Nominal size of aggregates			
	20 mm	14 mm	10 mm	6 mm
28 mm	100	-	-	-
20 mm	85-100	100	-	-
14 mm	0-40	85-100	100	-
10 mm	0-7	0-40	85-100	100
6.3 mm	-	0-7	0-35	85-100
5 mm	-	-	0-10	-
3.35 mm	-	-	-	0-35
2.36 mm	0-3	0-3	0-3	0-10
600 μm	-	-	-	0-2
75 μm	0-2	0-2	0-2	0-2

TABLE 11
Requirements for grading for chippings for hot rolled asphalt (BS 594:1985)

BS test sieve	Percentage by mass passing BS test sieve Nominal size of aggregates	
	20 mm	14 mm
28 mm	100	-
20 mm	90-100	100
14 mm	0-25	90-100
10 mm	0-4	0-25
6.3 mm	-	0-4
75 μm	0-2	0-2

(i.e. on higher speed roads), not less than 75 percent by mass of the chippings must be of the specified size.

The Department of Transport's "Specification for highway works" (Department of Transport, 1988) requires that chippings for surface dressings comply with BS 63 and should be "clean, hard and durable". Polished-stone value and aggregate abrasion value limits are also prescribed (See Chapter 5 and 6). The Department's specification also requires that pre-coated chippings for application to pre-mixed surfacings must be in accordance with BS 594:Part 1.

3.1.3.4 Surface dressing on concrete

On bituminous roads, surface dressing affords a cheap and rapid method of maintaining skid resistance by the application of aggregate of high resistance to polishing. On worn concrete roads texture depth is commonly provided by cutting transverse grooves into the hardened concrete surface, but even with specially designed machinery this technique is slow and costly. Surface dressing provides an alternative. The technique had been successfully used during the 1939-1945 war for a different purpose - to disguise a concrete road in Kent that was being used by German bombers to guide them to London.

Clemmer was one of the first to report that it is necessary to use aggregates of a higher quality when surface dressing concrete rather than a flexible material (Clemmer, 1943). More recently the problem was the subject of further study (Wright, N, 1976). Results from five full-scale road experiments carried out between 1969 and 1975 were analysed. They demonstrated the feasibility of applying single surface dressings to concrete roads carrying traffic in Category 1 (over 2000 commercial vehicles a day in one traffic lane in one direction). The preferred chipping size was found to be 10 mm. The problems of loss of texture depth by embedment of chippings, often associated with bituminous surfacings, did not arise when surface dressing these hard unyielding concrete roads. But aggregates used for surface dressing high-speed concrete roads were found to require greater strength in order to resist wear under the action of heavy traffic to maintain their rugosity. Recommendations were made for the maximum permissible aggregate abrasion value for chippings selected for surface dressing on concrete: like Clemmer, Wright found that they needed to be more stringent than those applicable to bituminous surfacings carrying similar traffic loads.

3.1.4 Resin-bound surfacings

Following success by American workers with synthetic resins, James carried out a trial in 1959 on Trunk Road A4 (James, 1960; James and Lamb, 1974). An area of the Colnbrook By-pass was surface-dressed using an extended epoxy-resin mixture with 1/8 inch to 1/16 inch (3 mm to 1.5 mm) grit. A range of five types of natural roadstone grit was compared with a range of seven types of abrasive grit, which included a refractory grade of calcined bauxite. It was concluded that the method would be satisfactory with normal roadstones, and that a very high skid-resistance could be obtained with a refractory grade calcined bauxite (see Fig. 7).

The specification for aggregates for use in resin-bound surfacings has been largely a matter of agreement between the few suppliers and their customers. However the Department of Transport have published a specification (Department of Transport, 1986) entitled "Resin-based high skid resistant surface treatment" which requires the aggregate to be calcined bauxite, that it should be clean and free from foreign matter, and that not more than 5 per cent is retained on a 3.55 mm British Standard test sieve and not more than 5 per cent is passed by a 1.18 mm British Standard test sieve.

Further information on resin-bound and other skid-resistant surfacings is give in Chapter 5.3.1.

3.1.5 Setts and pavers

A few roads are still surfaced with stone setts for ornamental purposes. These are mainly second-hand setts recovered from city streets supplemented by imports from Portugal. The appropriate British Standards specification is BS 435:1975 "Specification for dressed natural stone kerbs, channels, quadrants and setts". The manufacture of setts and other dressed stone products is a

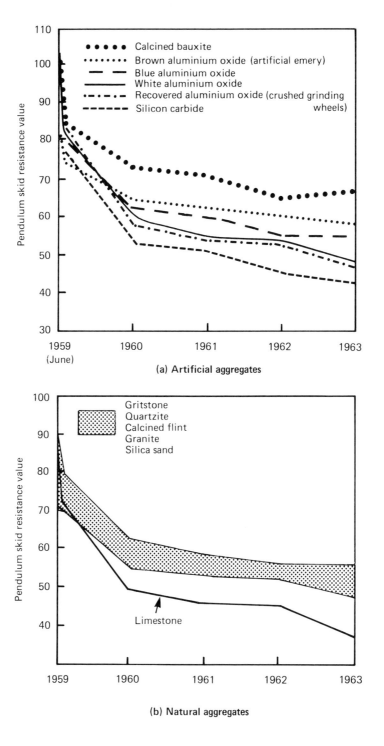

Fig. 7 Epoxy resin surface dressings, A4 Colnbrook By-pass, 1959.

skilled craft and has largely died out in Great Britain. However the Author has been led to understand that there is a renewed interest in the craft.

Clay and calcium silicate pavers are now more commonly used where an ornamental surfacing is desired. BS 6677:1986 "Clay and calcium silicate pavers for flexible pavements" sets out requirements for this type of surfacing.

A test has been developed for assessing the polish-resistance of pavers. This has been published as a draft for development by the British Standards Institution (DD 155:1986 "Method of determination of polished-paver values of pavers"), results are reported as polished-paver values (PPVs).

3.2 International recommendations for aggregates testing

This Section summarizes international reviews of the testing of aggregates that have been made during recent years. These were carried out with the object of assisting international standardisation.

The Permanent International Association of Road Congresses PIARC have reviewed the tests being used in member countries and have published recommendations (PIARC, 1979; PIARC, 1983). This work has been done in co-operation with RILEM and the American Society for Testing Materials (ASTM). A summary is given below:

3.2.1 Particle size distribution by sieving

Square-mesh sieves are recommended, but for the few countries using round-hole sieves the following conversion formula is given:

Size of square-mesh aperture = 0.8 x round-hole diameter

Two series of sieves are recommended, those of the International Standards Organisation (ISO) and ASTM. It is pointed out that since the particle size distribution results are usually converted to graphical form, the actual sizes are not important. However the 0.075 mm (75 µm) test sieve is recommended for separating sand from "filler".

Minimum masses for the test sample are given. Expressed in grams they are: 200 times the maximum size of aggregate for sizes equal or less than 20 mm, and 600 times the maximum size for aggregates greater than 20 mm. The choice of hand or machine sieving is optional and wet sieving is recommended when the aggregate is dusty or contains clay. The recommendation is to sieve dry if clean. If not clean, separate the <0.075 mm material by wet sieving. The end point is defined as when the mass passing through a sieve during one minute is less than 1% of the mass retained on that sieve.

3.2.2 Los Angeles test

The method recommended is that standardised by the American Society for Testing Materials (ASTM C131 and ASTM C535). It was found that few countries deviated from this procedure and any deviations were small.

3.2.3 Sand equivalent

The standard American Society for Testing Materials method is recommended (ASTM D2419) with mechanical shaking.

3.2.4 Polished-stone value test

The British Standard method of BS 812:1975 is recommended.

3.2.5 Quantity of material passing a 0.075 mm sieve

The recommended method assesses the quantity of fine material such as clay, mud, silt and dust. It is similar to that of BS 812:Part 103. The minimum mass of test sample is the same as recommended for the particle size distribution test. The sample is immersed in water until clay lumps disintegrate (boiling for 10 minutes if necessary), and the result is expressed to the nearest 0.1%.

3.2.6 Density measurements

The relative density is defined as the ratio of the mass of the solid matter of the particles to its volume excluding the volume of pores/voids within the particles. (*Author's Note:* the term "particle density" is currently being used in the place of relative density.) It is recommended that measurements be made on ground material all of which should pass a 0.075 mm test sieve.

The apparent relative density (now apparent particle density) is defined as the ratio of the mass of the solid matter of the particles to its volume including the volume of pores/voids within the particles.

The American Society for Testing Materials procedures are recommended (ASTM C127 and ASTM C128). These are carried out on a saturated surface dry sample using a pycnometer for material passing a 5 mm sieve, and a hydrostatic balance for larger material.

The bulk density is defined as the ratio of the mass of the solid matter of the particles to the overall volume of the particles including the volume of pores/voids within the particles and the voids between them.

For compacted bulk density measurements the amount of rodding, or jarring by dropping, is scaled to the size of the container.

The water absorption is defined as the ratio of the mass of absorbed water to the dry mass of the aggregate. This test is usually combined with the assessment of apparent relative density.

3.2.7 Aggregate shape

Length (L) is defined as the greatest distance between two parallel planes, each tangential to the particle.

Thickness (E) is defined as the width of the narrowest slot through which the particle can pass.

Width (G) is defined as the smallest square aperture through which the particle can pass.

From these three measurements the following coefficients may be derived.

Shape coefficient = L/E
Elongation coefficient = L/G
Flakiness coefficient = G/E

Shape and flakiness measurements are variously defined in different countries. Consequently it is recommended that the shape coefficient is measured by using calipers and the flakiness coefficient is measured by using slotted sieves.

The recommended calipers have two sets of jaws, the separation of one set being maintained (automatically) at one third of the other. The minimum masses for test samples are the same as are recommended for the particle size determination, with material less than 4 mm in size being discarded. The result is expressed as shape index which is the percentage by mass of the particles passing the smaller jaw, when set by the larger jaw.

The flakiness index is determined in a way similar to BS 812. However the individual sub-sizes are defined by fractions having a ratio of 1.25 (i.e. smaller than the ratio used in BS 812). An appropriate set of slotted sieves is defined.

3.2.8 Sample reduction to provide the test sample

The recommended sampling procedures are those of BS 812:1975, ISO/DIS 4847 and NFP18-553. Recommended sample reduction methods are also contained in these three standards.

3.2.9 Sensitivity to freezing

Two basic procedures have been employed in different countries. One assesses the change in granularity produced by a number of freezing and thawing cycles. The other measures the change in some mechanical property during these cycles.

The second method was considered to be more effective in showing up any micro-cracking that may have been produced, but the first method had the advantages of simplicity and being able to be carried out on site. Accordingly it was recommended that the test sample should be sieved through a 1.6 mm sieve before and after the freezing and thawing cycles, and then subjected to a Los Angeles test. This provides both measures of the effect of freezing and thawing.

3.2.10 Aggregate identification information

It is recommended that petrographic analysis should be carried out by a competent petrographer in the way that is defined in standards such as ASTM C295, BS 812:1975 and AFNOR P18-557. The method should include the following:

3.2.10.1 Mineralogical composition

Examination should be made to identify the minerals that may have an inferior bond with the added binder, or react with it in the long term, causing expansion and destroying the bonds previously established. For hydraulic binders these include, opal, chalcedony, crypto-crystalline quartz, sulphides and sulphates (such as pyrite, anhydrite and gypsum), some weathered dolomites and serpentines, biotite, muscovite and clay minerals. With bituminous binders the presence of certain minerals with high silica content, can imply a risk of stripping.

3.2.10.2 Degree of weathering

The exposure of some minerals to the action of water and air may lead to progressive loss of their mechanical properties.

3.2.10.3 Structure, porosity and soundness

These factors may have a significant influence on the mechanical properties and on the behaviour of certain rocks when exposed to the action of water or weather. The presence of clay between quartz grains can cause swelling, and high porosity can produce a low mechanical resistance and frost sensitivity.

3.2.11 Flow value of sands

The purpose of such a test is to give an indication of the surface texture and of the angularity of sand particles. These have an important influence on the resistance to plastic deformation of bituminous mixtures, and on the workability of bituminous and concrete mixes. The method recommended is that in the French standard AFNOR NFP 18564/1980. The principle of measurement is to time, in seconds, the flow through a calibrated funnel of a mass of sand. It is emphasised that the results of the test can be misleading if flat particles of mica are present in a round-grained sand.

3.2.12 Resistance to abrasion

Two methods are recommended, the French micro-Deval test and the aggregate abrasion test. The Los Angeles test is not included because it is considered to be mainly an impact test.

Although the micro-Deval test is simpler, the aggregate abrasion test is considered to indicate more accurately the resistance of an aggregate to the wear, especially when studded tyres are used.

The micro-Deval test can be carried out on either a dry or a wet aggregate; the wet aggregate is mostly used because, under field conditions, moisture is present. The recommended procedure is according to the French standard NFP 18-572/1978.

The recommended aggregate abrasion test is that of BS 812:1975 except that local silica sand may be employed as the abrasive.

3.2.13 Methylene blue test

The purpose of this test is to give a general indication of the presence and quantity of swelling clay minerals present in sands and fillers. The principle of the method is to add quantities of a standard aqueous solution of the dye to a sample until adsorption of the dye ceases. The blue value expresses the quantity of dye required to cover the total surface of the clay fraction of the sample with a mono-molecular layer of the dye. Thus the value is proportional to the product of the clay content times the specific surface of the clay. Typical values of specific surface (in sq.m./g.) are: montmorillonite (800), vermiculite (200), illite (40-60), kaolinite (5-20) and non-clay fines (1-3).

3.2.14 Compacted filler voids determination

The Rigden test is recommended to measure the void content of a filler in dry compacted state and thus, indirectly, the binder carrying capacity of the filler. The recommended procedure is according to BS 812:1975 or the Swiss standard SNV 670840/1977.

3.2.15 Stripping of bituminous binders

Two test methods are recommended, the static total water-immersion test (TWIT) and the loss of Marshall stability after immersion test.

The TWIT test applies to aggregates to be used for surface dressing. Coated aggregate is immersed in water and maintained at 40 degrees C. for 48 hours. After specified periods of immersion (1, 3, 24 and 48 hours) the proportion of the surface of the aggregate from which the binder has been lost is estimated visually.

The Marshall stability after an immersion test evaluates the effect of water on a bituminous mix by measuring the Marshall stability after immersion in water at 25 degrees C. for 7 days. The recommended procedure is that specified in ASTM D1559, except that at least 6 specimens are required.

3.2.16 Precision and accuracy

Differentiation is made between precision and accuracy and they are defined as follows:

The precision of a measurement process refers to the degree of mutual agreement between individual measurements.

The accuracy refers to the degree of agreement between individual measurements and the true value of the property of the material measured.

The definitions of reproducibility (R) and repeatability (r) are based on the International Standards ISO 3534 and ISO 5725.

The PIARC committee recommends that the value chosen for a specification should take into account the reproducibility of the test method. They refer to "sensible limits" that have been established from the reproducibility for some standards. They give as an example, "for a double limit (A1 and A2) the specification range shall not be less than four times the reproducibility (R), i.e. $A2-A1 > 4R$".

The PIARC review of test methods used by different countries has shown that there are three categories:

i   Test methods which include precision data, where the precision is acceptable.
ii   Test methods which include precision data, but where the precision has been incorrectly determined, or is unacceptably large.
iii   Test methods which do not include precision data.

The report underlines the problems associated with the precision of current test methods saying, "The third category, where no precision data is given, contains a large proportion of today's tests for road materials. The absence of precision data may be due either to their not having been determined, or their having been found to be so bad that it was decided not to publish them in the standard".

4 Tests for roadstones and aggregates

4.1 Test procedures

4.1.1 General

4.1.1.1 Uses of tests

A wide range of procedures for the testing of roadstones and aggregates have been specified in different countries at various times. Early British tests were mainly carried out on carefully selected and prepared individual specimens of rock with the object of determining the suitability of a particular rock as a source of road-making material. They were also suitable for testing shaped blocks of stone such as setts, which were widely used for road surfacing at the time the tests were adopted. Because of the considerable variation of rock within most quarries and the unsuitability of the earlier tests for crushed and screened aggregates, they were superseded by tests on samples of aggregates rather than solid rock. These had the additional advantage that they take into account any variations in properties that might result from differences in particle shape that can arise from different techniques in crushing: for example, the strength of an aggregate can be significantly decreased if it contains a high proportion of flaky or elongated particles.

The choice of test method will depend on the need. Some of the main needs are as follows:

a.   Checking against specification requirements

Limits are laid down in many specifications and require tests to be carried out in a particular way (usually to a national standard). Such tests need to be closely specified, be accurate and have established precision statements.

b.   Monitoring production

Some tests are used to monitor the production from a source to ensure that it complies with specification limits. Because of the correlation that often exists between different tests, particularly within a class of aggregates, the tests used may not be the same as those in the specification. This can save both time and expense.

c.   To assess a potential new source

Some tests are particularly applicable to the source material of, say, a new quarry. The crushing strength test is an example; the aggregate crushing test is less appropriate because results would be affected by grading and shape. Other tests are quite inappropriate (shape is an example) because they are usually more dependent on the processing plant than on the basic material.

d.   Screening tests

These tests may be employed to make a quick and inexpensive check on quality. The result will indicate whether further testing is necessary. With tests of long duration or of considerable expense a preliminary screening test is particularly valuable. Such a test is the methylene blue dye test that takes only 24 hours and can eliminate the need for lengthy drying shrinkage tests on mortar prisms.

e.   As a reference datum

Test procedures that might require special equipment or be of long duration might, on some occasions, be of value in checking the accuracy of a more generally used test. The saturated-air water absorption test is such a procedure that can be used to check the BS 812 water absorption test.

f.   Research purposes

Many tests that have been developed for research purposes are too expensive in time and/or material to be of general use. However many of the commonly used tests were first developed as research tools.

4.1.1.2 Precision

a.   Background

The importance of the reproducibility of tests results was underlined in Road Research Special Report No. 3 (Phemister *et al*, 1946). This Report included estimates of the reproducibility for the seven commonly used mechanical tests of the time, in terms of coefficient of variation (standard deviation expressed as a percentage of mean value). The estimates (see Table 12) show that the results of the tests on aggregates are much more reproducible than the tests made on single pieces of rock. It was found that one aggregate crushing or Los Angeles test gave the same probability of a true result as 90 Page impact tests or 60 crushing strength tests.

TABLE 12
Reproducibility of mechanical tests on roadstone

Test	Coefficient of variation	Number of samples that must be tested to ensure 0.9 probability that the mean will be:	
		within ±3% of true mean	within ±10% of true mean
Dry attrition value	5.7	10	1
Wet attrition value	5.6	9	1
Abrasion value	9.7	28	3
Impact value	17.1	90	8
Crushing strength	14.3	60	6
Aggregate crushing value	1.8	1	-
Los Angeles abrasion value	1.6	1	-

b. *Repeatability and reproducibility.*

A large number of factors can have a bearing on the variability of the results of tests on aggregates. Some arise in the production of the aggregate, some in the sampling methods used and others in the testing procedure itself.

There are two main features of the variability of testing, repeatability (r) and reproducibility (R). Which, respectively give a measure of the degree to which the results of two tests on identical samples are repeatable when tested at the same and different laboratories. Knowledge of these values gives a means of detecting faulty testing.

c. *British Standards and precision*

The British Standard specification for the testing of aggregates has published a guide (BS 812:Part 101:1984) to sampling and testing aggregates, which, apart from general guidance, gives procedures for assessing the precision of tests in the other Parts to the Standard. This Standard is based on BS 5497, the British Standard for precision assessment, interpreted according to the special needs of aggregates. Five measures of precision have been defined to meet the various needs of testing.

> Reproducibility (R) is defined as "The value below which the absolute difference between two single test results obtained with the same method on identical test material under different conditions (different operators, different apparatus, different laboratories and/or different time) may be expected to lie with a probability of 95%."

> Repeatability (r) is similarly defined but is concerned with the "same conditions (same operator, same apparatus, same laboratory and a short interval of time)".

In addition there are two variations of reproducibility (R1 and R2) and one of repeatability (r1). These are more appropriate to the practical conditions of testing. R1 and r1 relate to "different test portions of the same laboratory sample" rather than "identical test material", and R2 relates to "different laboratory samples from the same batch".

d. *Uses of precision estimates*

r is of limited value as it can only be estimated where the test portion is not changed by the test method. r1 does not suffer from this restriction but will include any sample reduction error. Estimates of r and r1 can be used in two ways.

i   To screen data when tests are repeated within a reasonable time. In this event pairs of tests are rejected if the difference between them exceeds the estimates for the test concerned.

ii To monitor the testing of a laboratory. In this case the comparison of values obtained from 20 pairs of tests is considered to be sufficient.

R1 is of value in comparing tests by two laboratories where test portions from the same laboratory sample are used. However if different samples are used the comparison must be made with R2.

If sampling presents a problem (e.g. where there is segregation in stockpiles) the sampling error variance should be taken into account.

e.   Specification limits

BS 812:Part 101 states that the achievable precision of a test method should be taken into consideration when fixing specification limits using R2, which includes allowances for sampling errors and sampling reduction errors.

4.1.2   Sampling

4.1.2.1 Variability in the quarry

Samples of aggregate taken at different times or from different parts of the same quarry can show considerable differences in properties. An investigation (Director of Road Research, 1960) was made made to determine the extent to which different samples of roadstone from a number of sources vary in their resistance to crushing, abrasion and polishing. Five quarries - three producing gritty limestones, one producing igneous rock and another producing a tough indurated breccia resembling igneous rock - were studied: in four of them a large number of samples were taken from different places distributed over the working faces; in the fifth this was not possible and the samples were taken from the output conveyor. The samples were subjected to tests for resistance to crushing (10% fines test), abrasion (aggregate abrasion test) and polishing (polished-stone coefficient test), and a statistical analysis of the results was made.

Table 13 gives a summary of the results of this statistical analysis, and Fig. 8 shows the distribution of the values of polished-stone coefficient (approximately polished-stone value x 0.011) for each quarry. The results indicate the extent of the variability that can occur in samples from the same source, even with the strong and apparently uniform rocks. The results show clearly that, where the attainment of a maximum requirement is critical, as with for instance the polishing resistance, it is important that samples should be taken from the finished products of current production, and at intervals from the material actually to be used in the road surfacing.

4.1.2.2 Variability of the finished aggregates

By 1962 road engineers were making increased use of test results as a basis for selecting road aggregates, and because of the inherent variability in the properties of roadstones it became even more important that the results should provide a reliable indication of the properties of the materials supplied. It was found that where the behaviour of the aggregate in the road differed significantly from that expected from tests results it was usually traceable to the fact that single samples taken either at the quarry face or from the finished aggregate were not representative of the material used in the road. The problem was therefore studied with the object of developing a more satisfactory sampling procedure (Shergold, 1963).

A large number of samples were taken from the rock face and from the finished aggregate at seven quarries, covering a range of plants and types of rock. The samples were subjected to the British Standard tests for determining their resistance to crushing and polishing (BS 812:1960). The

TABLE 13
Variability in the physical properties of roadstones from five quarries

Type of stone	Number of samples tested	Mean results	Range of results	Standard deviation
Polished-stone coefficient:				
Gritty limestone A	50	0.63	0.50-0.75	0.047
Gritty limestone B	50	0.61	0.53-0.68	0.040
Gritty limestone C	45	0.60	0.53-0.68	0.035
Porphyrite-breccia	30	0.70	0.60-0.81	0.048
Granophyric diorite	30	0.63	0.59-0.69	0.024
Ten-per-cent fines value:				
Gritty limestone A	50	21	17-30	2.3
Gritty limestone B	50	13	10-16	1.4
Gritty limestone C	36	15	12-18	1.9
Porphyrite-breccia	30	27	12-38	5.9
Granophyric diorite	30	32	27-37	2.6
Aggregate abrasion value:				
Gritty limestone A	10	11	8-13	1.3
Gritty limestone B	10	26	14-36	5.7
Gritty limestone C	10	19	13-23	3.3
Porphyrite-breccia	10	5.5	2.8-10.4	2.2
Granophyric diorite	11	4.0	2.9-5.4	0.7

variability of the test results was then determined in terms of their coefficient of variation. The main findings were:

i   Samples for test for road engineering purposes should always be taken from the finished aggregates, and never from the quarry face.

ii A test result that is reliably representative of a whole day's output of the finished aggregate could be obtained by sampling at hourly intervals over a period of a day, thoroughly mixing these samples, and reducing the composite sample by means of a sample divider to the quantity required for test.

These findings were later incorporated into revisions of BS 812.

4.1.2.3 Variability of blastfurnace slag

A collaborative study by the British Slag Federation and the Road Research Laboratory (Hosking, 1967a) of the blastfurnace slags being produced showed that the variability between hourly samples of the production was relatively small; daily samples showed a greater variability, and the greatest variation occurred between samples taken monthly.

4.1.2.4 Sampling for sieve analysis

The British Standard method of carrying out sieve analyses (BS 812) has required the tests to be carried out on one sample only, unlike the two (or more) required for other tests. This,

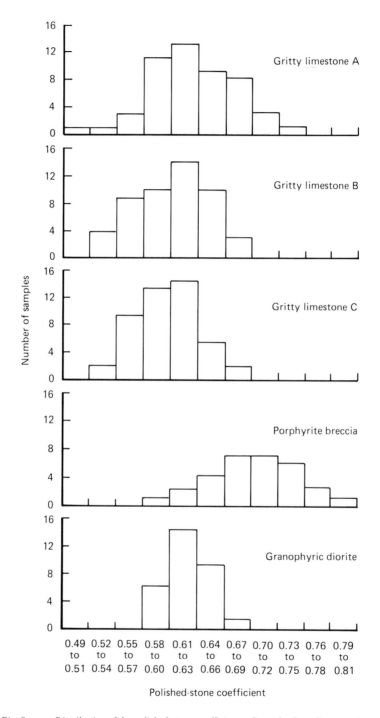

Fig. 8 Distribution of the polished-stone coefficients of samples from five quarries.

TABLE 14
Estimates of standard deviations of set of gradings

		Standard deviations of proportions finer than BS sieve - mm							
Ref	*37.5*	*20*	*10*	*5.0*	*2.36*	*1.18*	*0.60*	*0.30*	*0.15*	*0.0075*
101	-	3.9	3.7	3.0	2.4	1.7	1.3	0.9	0.7	0.5
7	-	1.5	1.5	1.7	1.3	1.1	1.0	1.1	0.9	0.9
103	-	4.3	3.0	2.6	2.0	1.7	1.5	0.9	0.7	0.7
104	-	3.6	3.8	3.1	2.8	2.5	2.2	1.1	0.4	0.3
105	-	3.9	3.8	3.3	2.9	1.7	1.5	1.3	1.2	1.1
106	-	3.2	3.4	3.5	3.1	2.8	2.5	2.0	1.4	1.2
107	-	8.4	7.2	5.2	3.9	3.2	2.7	2.1	1.7	1.4
108	-	4.5	3.6	2.8	2.2	1.9	1.6	1.4	1.2	1.1

presumably, has been because it was a quick and cheap test which has been considered adequate for the purposes concerned. Pike became aware that sampling and testing errors could contribute markedly to the variability of grading results and made a study of the problem (Pike, 1979).

Eight sub-base materials (maximum size 37.5 mm) and two concrete aggregates (a 20 mm to 5 mm gravel and a sand) were subjected to the standard (BS 812:1975) sample-reduction procedures, except that all sub-fractions were tested.

It was found that a range of 6 per cent in grading at the 5 mm size was introduced by riffling the sub-base materials to the appropriate size for testing. A similar result was found at the 10 mm size for the 20 to 5 mm gravel, and a range of 4 per cent was found at the 600 µm size for the concreting sand.

Pike pointed out that these (and other) variables need to be taken into account when specifications are being drafted.

Details of the variability are shown in Table 14.

4.1.2.5 British Standard specifications

BS 812:Part 101:1984 gave information on precision estimates and the way they are used to improve testing, and BS 812:Part 102:1990 gives practical information on the sampling methods to be used for various types of aggregate. The methods specified are based on the general principles given in BS 5309:Parts 1 and 4. It is recommended that materials finer than 75 µm should be sampled according to BS 1550:Part 1.

BS 812 defines batch (the material being sampled), sample increment, bulk sample, laboratory sample and test portion (the material actually tested) and specifies methods that have the object of obtaining a representative test portion from the batch. These include quartering and the use of a sample divider.

An appendix to BS 812:Part:101:1984 gave definitions of a number of rock types, a specimen sampling certificate and information of value in assessing shape and surface texture.

4.1.2.6 Sampling for particular tests

Some tests are carried out on an aggregate test portion of a particle size which may not be the same as the material being used. The polished-stone value test is an example. Others such as the aggregate crushing test can be carried out on a range of sizes, but tests are normally carried out on the 14 mm to 10 mm fraction. Where this occurs particular care must be taken to ensure a representative sample. Laboratory crushing can affect strength measurements by changing the shape of the material and sieving out a minority fraction to obtain enough sample to test can also lead to considerable error.

4.1.3 Tests for strength

4.1.3.1 Lovegrove attrition test

This, the first important British roadstone test, is included because of its historic value and also to show the way in which both tests and specifications for aggregates have changed during this century.

Lovegrove's attrition test (Lovegrove *et al*, 1929), sometimes referred to as the Hornsey test, was the first attrition test to be used on a large scale in Great Britain. The test apparatus consisted of three rotating cast-iron horizontal cylinders, driven by a gas engine through a counter-shaft and bevel gearing, enabling three samples to be tested simultaneously. The cylinders were 11 inch (Imperial units have been reproduced in these early test descriptions) internal diameter, with three 1 inch by 1 inch angle-iron ribs, bolted length-wise in the inside at equal distances apart and parallel to the axis of rotation. Samples were broken to a 2 inch gauge, as supplied for road work, numbering about sixteen stones and weighing 4 lb. The number of revolutions, recorded by a counter, was confined to 8,000 and the speed to twenty revolutions per minute. For wet tests about half a gallon of water was placed in the cylinder. The percentage weight of chips (not exceeding 1/4 ounce apiece) and dust formed gave the attrition value. After the test the stones had an appearance of rounded water-worn pebbles.

The ribbed horizontal cylinder effected a combination of attrition and slight impact, whereas the Deval attrition test effected attrition only and the Page impact test impact alone.

Early in the 20th century the Lovegrove attrition test was discontinued in favour of the Deval attrition test.

4.1.3.2 Deval attrition test

a.   Original test

M. Deval, a French engineer, designed an attrition machine that was similar to Lovegrove's machine. It consisted of an iron cylinder 8 inches in diameter and 13 1/2 inches deep, mounted diagonally on a rotating axle. 11 lb weight of stone, broken to pass through a 2 1/2 inch ring, were placed in the cylinder, which was then closed and rotated at a speed of 30 revolutions per minute for a period of five hours. The contents, consisting of the abraded fragments of stone, and all particles larger than 1/16 inch were again weighed, the difference representing the dust formed by abrasion. The result was expressed as coefficient of wear which was calculated by dividing 400 by the mass of dust formed.

b.   British Standard test

A form of the Deval attrition test superseded the Lovegrove attrition test in Great Britain and was included in the appropriate British Standard, BS 812. In this test 5 kg of 2 inch to 1 1/2 inch (50 mm to 37.5 mm) stone was placed in the inclined cylinder which was rotated at 30 rpm for 10,000 revolutions. The dust formed was then sieved out on a No. 10 (about 1.5 mm) BS sieve and weighed. The British test was carried out both dry and wet (when one gallon (4.5 litre) of water was added to the charge). The results were not expressed as coefficient of wear, as in France, but were expressed as "dry attrition value" or "wet attrition value" respectively, these being the amount of dust formed expressed as a percentage of the dry mass. Lower values indicated a better material. The method was similar to test method ASTM D2-33, published by the American Society for Testing Materials, but discontinued in 1972.

BS 812:1938 included details of versions of the test for 1 inch to 3/4 inch (25 mm to 19 mm) and 1/2 inch to 3/8 inch (12.5 mm to 9.5 mm) stone. Factors of 1.5 and 2.5 respectively were used to give resulting values that were roughly equivalent to those obtained for 2 inch stone.

The Deval attrition test has not been used in Great Britain for the testing of road aggregates for many years. The last British Standard specification for this test was in BS 812:1951, when only the 2 inch stone version was given. The attrition tests for 1 inch and 1/2 inch being omitted because "they give results of doubtful significance". The Deval attrition test was omitted altogether from BS 812:1960 as being "of doubtful value". Nevertheless it has continued to be used for testing railway ballast, where a wet attrition value of not greater than 6 is required for main lines, and a value of 8 for other lines.

c.   Modified Deval attrition test

Modifications, other than by the use factors, were made by the Road Research Laboratory in order to be able to test chipping sizes. When the size of the material in the charge was reduced, the percentage of fines formed fell correspondingly, but by using a suitable factor it was found possible to compare the results with those for the 2 inch material. The amount of dust formed was, however, quite small even for 2 inch material and attempts were made to increase it by including steel balls in the charge. Preliminary work was done with a charge of 6 steel balls weighing 3 kg added to a sample of 5 kg of 1/2 inch to 3/8 inch chippings, a 10 mesh (No. 10) sieve being used to separate the fines. This modified test was not adopted by the British Standards Institution and has not been generally used in Great Britain.

4.1.3.3 The micro-Deval test

This test can be carried out on either a dry or a wet aggregate. A sample of standard-size (14/10, 10/6 or 6/4 mm.) is placed with a charge of 10 mm diameter steel balls in a stainless-steel cylinder. The cylinder is rotated about its longitudinal axis and the result of the test is expressed as the percentage by mass of the material that is reduced to pass a 1.6 mm test sieve. The recommended procedure is according to the French standard NFP 18-572/1978.

4.1.3.4 Los Angeles abrasion test

The Deval attrition test takes over five hours to perform. The American Los Angeles abrasion test (sometimes referred to as the Los Angeles rattler test) takes only 20 minutes to achieve comparable results by introducing some twelve 1 7/8 inch steel balls in a steel cylinder fitted with an internal shelf and rotated at 30 rpm for 500 revolutions. Two alternative gradings, 1 1/2 inch down and 3/4 inch down, can be tested. The result of the test is expressed as the percentage by mass of material passing a No.12 ASTM sieve (equivalent to a No.10 BS sieve) after test. Two versions of the test are specified in the ASTM designations, one for coarse and the other for fine aggregates (ASTM C131 and ASTM C535 respectively). Good results were reported in the 1939 Annual Report of the Road Research Laboratory. Results in Britain at this time were known as "Los Angeles values", the lower the value the stronger the aggregate.

The Los Angeles abrasion test is used in many parts of the world. It has been seldom used in Great Britain where the aggregate crushing test (and later the 10% fines test) have been employed for the same purpose. This choice may have been influenced by the torturous sound emitted whilst the test is running, but a suitable sound-proof cabinet will reduce this to an acceptable level.

The Los Angeles abrasion test values have been the subject of extensive comparisons with in-service behaviour (Clemmer, 1943). These included direct quantitative measurement of crushing under the roller, measurement of the crushing strength of concrete made from tested aggregates, and reports of road usage from various States in the USA in relation to surface dressing, bituminous surfacings and concrete. It was concluded that the test gave a satisfactory indication of the in-service behaviour of an aggregate. Suggested maximum Los Angeles abrasion values were 40 for bituminous materials and 50 for concrete aggregates.

During the development of the aggregate crushing test comparisons of results were made between the aggregate crushing values and Los Angeles abrasion values of a range of roadstones (Markwick and Shergold, 1945). The good correlation obtained is shown in Chapter 4.2 below.

4.1.3.5 Crushing strength test

Test specimens are prepared in the form of cylinders (25 mm diameter, 25 mm length) by drilling and grinding a suitable piece of stone. Three specimens are usually tested during a single determination, but four are tested if the rock has planes of weakness (when two specimens should have such planes at right angles to the axis of the cylinder). The test is carried out by crushing the specimens in a compression testing machine, spherical seatings being used to ensure axial loading of the specimen. The results are reported as the crushing strength in MN/sq.m. (formerly lb./sq.in.). Higher results are given by stronger materials.

The crushing strength test was the only test that employed shaped specimens to be retained in the British Standards specifications until recent years. It was discontinued in 1975. Its particular value was in assessing samples from new rock formations (particularly samples obtained with a core-drill) and for comparing rocks independently of any differences that might arise from crushing and screening.

The requirement for sandstone kerbs, channels, quadrants, flags and setts used to be a minimum of 15,500 lb./sq.in. (100 MN/sq.m.). Most road-surfacing aggregates have crushing strengths of more than 200 MN/sq.m.

4.1.3.6 Aggregate crushing test

a.   Standard test

A cylinder crushing test was adopted as a German national standard in 1935. The British aggregate crushing test was a slightly modified form of this test, the results being expressed as the percentage of material passing a No. 7 (2.4 mm) British Standard test sieve rather than by the change in fineness modulus, as in Germany. This test was studied in detail (Markwick and Shergold, 1945), and the results of this work led to its adoption by the British Standards Institution in BS 812, where it remains to the present day. Resistance to crushing of a sample of aggregate is measured by submitting a 100 mm deep bed of 14 mm to 10 mm chippings in a nominal 150 mm diameter hardened-steel cylindrical mould to a load of 400 KN in a compression testing machine. The percentage by mass of fines passing a 2.80 mm test sieve formed in the test is known as the "aggregate crushing value" (ACV). For sizes other than 14 mm to 10 mm, tests may be made by using appropriate test sieves to define the fines, and a nominal 75 mm diameter mould may be used for the smaller sizes of aggregate.

The aggregate crushing test is considered to be a good general purpose test for measuring the strength of most aggregates. It not only takes account of any weakness that might arise from poor particle shape, but is suitable for testing crushed and screened aggregate both at the quarry or after delivery to the customer. Because of the large number of particles that constitute the test sample, reproducibility is much better than that of the crushing strength test. When tests on different sizes of aggregate are compared it is usual to find that the strength increases (smaller value) with decrease in size of the aggregates; this effect varies with different roadstones and is thought to be due to the elimination of planes of weakness during the crushing process. However where the source material is of widely ranging strength, the smaller aggregate sizes produced are likely to contain more of the weaker material.

Exceptionally strong aggregates yield values of 10 or less, and aggregate crushing values of 25 or less are produced by aggregates which are generally considered to be strong enough for most roadmaking purposes. Materials with values as high as 35 can be used in road surfacings if special precautions are taken (such as the use of rollers with rubber tyres), but such aggregates usually lack adequate abrasion resistance. Weaker aggregates may be used in concrete, for surfacing minor roads and in the lower pavement layers of bituminous roads. Because the aggregate crushing test is usually unreliable for aggregates yielding values in excess of 30, the 10% fines test is preferred for such materials.

b.   1/4-standard aggregate crushing test

The British Standard aggregate crushing test was found to be unsuitable for the weaker types of aggregate because the particles pack down to an almost solid mass in the early stages of loading. Results of a 10 tonne aggregate crushing test (generally known as the 1/4-standard crushing test or 1/4-s test) were found to agree more closely with road performance. Although his test showed considerable promise, it was abandoned in favour of the 10% fines test because the latter can be used to test aggregates of all strengths.

4.1.3.7 Ten-per-cent fines test

a.   Standard test

This test was developed (Shergold and Hosking, 1959) to overcome the problems with the aggregate crushing test when testing weaker materials. Resistance to crushing of an aggregate is measured by submitting a sample of chippings to an appropriate load in a compression testing machine in the same way as in the aggregate crushing test. The difference between the tests is that in the 10% fines test the load is adjusted to give 10% fines and the test result is reported in kN as the "ten-per-cent fines value" (10% fines value). This procedure ensures that different samples are crushed to the same extent and so overcomes the deficiencies of the aggregate crushing test, which is insensitive to differences between weak aggregates and yields anomalous results when used with weak porous aggregates such as light-weight blastfurnace slags.

The 10% test is equally suitable for the strongest and weakest aggregates and could ultimately replace the aggregate crushing test. It yields results which range from about 400 kN for the strongest aggregates down to 10 kN or less for weak materials such as crushed chalk and brick.

Unlike the aggregate crushing test, the metrication of standards has led to a numerical change in the reporting of test results. 10% fines values are now expressed in kilo-Newtons (kN) rather than tonnes force: 1 tonne force is almost exactly equivalent to 10 kN.

b.   Modified ten-per-cent fines test

Some aggregates yield significantly lower values (i.e. are weaker) when tested in a water-saturated state than in the dry state as normally tested. These should preferably be tested in the saturated state; experience suggests that a minimum (saturated) 10% fines value of 5 tonnes (50 kN) would be appropriate for unsurfaced roads, sub-bases and for the bases of more lightly trafficked roads (Hosking and Tubey 1969).

4.1.3.8 Page impact test

In the Page impact test the specimen of rock, 25 mm in diameter by 25 mm long, was subjected to automatically repeated blows from a falling tup weighing two kilogrammes. By means of a screw the fall was increased by 1 cm between each blow, until fracture occurred. The test result (impact value) was reported as the height in centimetres at which fracture occurred. This test was included in BS 812:1938, until the 1951 revision when it was superseded by the aggregate impact test which was considered to give more reliable results.

4.1.3.9 Aggregate impact test

a.   Standard test

In the mid-1940s the aggregate impact test was developed from a test originally devised by G Stewart of Cape Town University (Stewart impact test). Resistance to impact of a sample of aggregate is measured by subjecting a 28 mm deep bed of 14 mm to 10 mm chippings, in a 102 mm diameter hardened-steel cup, to 15 blows from a 14 kg hammer falling from a height of 380 mm. The percentage mass of fines (passing a 2.80 mm BS test sieve) formed in the test is known as the "aggregate impact value" (AIV). Higher values are given by weaker materials. For the majority of aggregates the aggregate impact value and the aggregate crushing value are numerically similar, but brittle materials such as flints yield aggregate impact values which are about 5 units higher than their aggregate crushing values.

Although the purpose of this test was to evaluate the brittleness of an aggregate, it is extensively used as an alternative to the aggregate crushing test. It requires a smaller test sample, is quicker to carry out and requires less expensive equipment (unless a compression test machine is already available). However the smaller test sample implies a poorer reproducibility and care must be taken to ensure that a sufficiently rigid base is available on which to carry out the test. The equipment is portable and so may be used at the quarry or road site, provided care is taken about the base.

b.   Modified aggregate impact test

Differences in value between water-saturated and dry samples have been recorded for some materials, as with other strength tests. If such differences are suspected, the standard test (which uses dry material) should be replaced by a procedure using a saturated sample. Further modification (Hosking and Tubey, 1969) has also been recommended for weak aggregates. Fewer blows of the hammer are employed and the results adjusted to allow for the reduction in fines so produced. Experience of such weaker materials suggests that a maximum (saturated) modified value of 40 would be appropriate for unsurfaced roads, sub-bases and for bases of more lightly surfaced roads.

4.1.3.10 Tensile strength tests

In view of the possibility that the tensile strength of the stone has an effect on the properties of concrete and bituminous surfacings, a study was made of the tensile strength of eleven roadmaking rocks, and the results compared with those obtained for the crushing strength of the rocks (Hosking, 1955). Because a direct tensile test would be unsatisfactory for testing rock, the flexural strength and indirect tensile strength were measured to study this property, as described below.

a.   Flexural strength test

Cylinders 25 mm in diameter by 125 mm in length were prepared by drilling and grinding a suitable piece of rock. These cylinders were subjected to third-point loading (i.e. the load is distributed equally between the two points that divide the supported length of the test cylinder into three equal parts) in suitable shackles in a compression testing machine until they failed (more uniform results are obtained than by central-point loading). The flexural strength was then calculated in kN/sq.m. by using a standard formula. This method was similar to that used for concrete (BS 1881:1952).

b.   Indirect tensile strength test

Test cylinders of 25 mm diameter and 25 mm length were used for this test; these were placed on their side in a compression testing machine and loaded until they split longitudinally. The indirect tensile strength was then calculated by using a standard formula. Specimens were prepared, wherever possible, from cores drilled in three mutually perpendicular directions from large rock specimens. The method is similar to that developed in Brazil for concrete.

Tensile strength tests are seldom used for roadstones, but they are more capable of detecting defects due to planes of weakness in the rock than the crushing strength test.

4.1.3.11 Fatigue strength tests

The Author carried out a series of tests on rock specimens in the mid 1950s with the object of determining whether a reduction in strength occurred both under repeated loading and under sustained static loading. The repeated load testing technique involved the preparation, by drilling and grinding, of test specimens of the Wöhler type. Rocks were chosen and specimens prepared so as to be as near identical as possible. The results (unpublished) showed that any fatigue effect was small (less than 10 per cent) and could not be rigorously separated from differences that could occur as a result of the variability of the tests. The static loading tests were flexural in nature and were carried out in a similar manner to the Wöhler tests, with similar results.

4.1.3.12 Dorry abrasion test

In the original Dorry abrasion test (also known as the Dorry hardness test) the specimens of rock, two in number, each measuring 25 mm in diameter by 25 mm long, were pressed against the surface of a cast steel disc, rotating in a horizontal plane, with a force of 250 grammes per square centimetre. Crushed quartz was used as an abrasive and was fed in through two small hoppers. The amount of material abraded was measured after 1,000 revolutions of the disc. With minor changes, the test was adopted by the British Standards Institution and incorporated in BS 812:1938. Leighton Buzzard silica sand was used as the abrasive and results were expressed as the coefficient of hardness (however results were often referred to as abrasion values). The coefficient of hardness being calculated by subtracting from 20, one third of the loss in mass in grams.

The Dorry abrasion test was superseded by the aggregate abrasion test in the 1951 revision of BS 812.

4.1.3.13 Aggregate abrasion test

The abrasion machine is similar to that used for the Dorry test and the test was developed in 1949. Resistance to abrasion is measured by finding the percentage loss in mass suffered by 33 cu.cm. of cubical (i.e. neither flaky nor elongated) 14 mm to 10 mm chippings when mounted in a single flat layer in a suitable setting medium and subjected to a standard abrasion procedure on a lap using a standard 600 µm to 425 µm Leighton Buzzard silica sand as the abrasive. The percentage loss in mass of the chippings is known as the "aggregate abrasion value" (AAV) and ranges from about 1 for hard flints to over 16 for aggregates that would normally be considered too soft for use in road surfacings.

Inadequate abrasion resistance of road-surfacing aggregates means an early loss of the texture depth required to maintain high-speed skidding resistance. Information on the limits introduced by the Department of Transport is given in Chapter 7.

4.1.4 Density and water absorption

Various measures of aggregate density have been used over the years. The Author has retained the expression used at the time for each subject under review.

The early British tests for specific gravity (eg BS 812:1943) were carried out on three stones each weighing about 50 grams by drying at 100 degrees to 110 degrees C. for 72 hours, weighing, immersing in water for 72 hours and then weighing in air and water. The specific gravity was calculated by dividing the dry mass in air by the apparent loss in mass of the stone in water. From these measurements the water absorption was also determined.

A later version of the test employed a sample of aggregate rather than large stones, and reduced the drying and soaking times, but problems of reproducibility were still encountered.

A study (Shergold, 1953a) was therefore made of the cause of the rather variable results obtained in the British Standard tests for the specific gravity and water absorption of coarse aggregate. It was found that the main causes of variability were attributable to the air entrapped in the aggregate when it was weighed under water and from differences in different observers' estimates of the saturated, surface-dry condition. It was also found that the aggregate could be more thoroughly saturated in the 24 hours specified, if it was soaked before drying instead of afterwards, but little advantage was gained by the use of reduced pressure or boiling. On the basis of this work, an improved method of test was drawn up that gave more reproducible results. Further refinement of this procedure led to the methods described below.

4.1.4.1 Tests for particle density and water absorption

Several methods of measuring the particle density of aggregates are specified which make use of measurements of the mass of the sample in air and in water. Either a wire basket, a gas jar or a pycnometer is employed to contain the sample, the choice being governed by the grading of the sample concerned. Particle density (which now replaces specific gravity) can be expressed on an oven-dried basis, on a saturated surface-dry basis or as an apparent particle density. The oven-dry figure is normally used for road engineering purposes.

Water absorption is normally obtained at the same time as the particle density; it is the difference in mass before and after drying the sample at 105 +/-5 degrees C. for 24 hours, expressed as a percentage of oven-dry mass.

The most recent specification for these tests is that given in BS 812:Part 107 "Methods of determination of particle density and water absorption".

Particle density is frequently used when an accurate assessment of the rate of spread of chippings to be used in a surface dressing is required.

Water absorption is important in the mix-design of concrete and may also give indications of the frost-susceptibility or other weakness of an aggregate. A high water-absorption (over 2 per cent) is common in aggregates that are frost susceptible. However it must be borne in mind that some aggregates (such as some blastfurnace slags) can have a high water absorption but are not frost susceptible. A high water absorption is also indicative of aggregates which tend to absorb binders.

A weakness of the methods in general use for measuring the water absorption of aggregates is the subjective assessment of the saturated surface dry condition. Work has been carried out at the University of Birmingham (Hughes and Bahramian, 1967) to develop an accurate laboratory test that would be suitable for research work and be able to provide a datum for other methods. This has led to the development of a saturated air test method that provided an accurate laboratory test for use on either continuously graded or single-sized aggregate. The method has given results on continuously graded fine aggregates that were numerically similar to the BS 812 test method.

4.1.4.2 Bulk density test

The results of bulk density tests on aggregates are used when it is required to convert mass to volume or *vice versa*. It is of particular importance when using lighter-weight blastfurnace slag aggregates. The effect of the grading and shape of an aggregate on its bulk density has been studied by the Road Research Laboratory (Moncrieff, 1953).

The British Standard test for the bulk density of aggregate (BS 812:1960) consisted of a determination of the mass of aggregate required to fill a container of 1/10, 1/2 or 1 cu.ft. (0.03, 0.015 and 0.003 cu.m.) capacity, with the aggregate either uncompacted or compacted by tamping in a prescribed manner. The size of the container was scaled to the size of the aggregate to be tested, but the same compactive effort was at this time required for all sizes of container, whereas it seemed more logical to scale the compactive effort to the size of the container. An investigation of this point (Hosking, 1961) showed that the British Standard method gave small but significant differences between the values for the same aggregate when tested in different sizes of container, and the results showed that more nearly equal results could be obtained if the amount of compaction was scaled to the area of the aggregate being tamped.

As part of this investigation, determinations were made of the bulk density of aggregates after they had been loaded into lorries at a crushed-stone quarry and at a gravel pit. The values obtained were slightly higher than the value of uncompacted bulk density obtained on the same aggregate in the laboratory test, the difference ranging from 1 lb/cu.ft. (16 kg/cu.m.) for single-sized aggregates to 14 lb.cu.ft. (224 kg/cu.m.) for a damp "all-in" aggregate. The results and conclusions from this investigation led to improved methods of carrying out the test.

Co-operative work by TRRL and ACMA has shown that the measurement of bulk density of either compacted sand or of asphalt aggregates can give a reliable method of determining the optimum binder content of the mix. This provides an alternative to the BS 594 design procedure.

The latest British specification for this test has been drafted as BS 812:Part 108 "Methods for determining the bulk density, optimum moisture content, voids and bulking". Four volumetric containers are specified (0.03, 0.015, 0.007 and 0.003 cu.m.) which relate to the appropriate particle size of the aggregate and the number of compactive blows needed. This draft standard also includes a vibrated bulk density test and a method of determining the bulk density of filler in kerosine.

4.1.5 Shape and texture

4.1.5.1 Roundness index

The terms "angular" and "rounded" have been frequently referred to in specifications and versions of BS 812 have included illustrations to demonstrate these and other shape factors (e.g. "irregular", "flaky", "elongated"). The Road Research Laboratory studied the problem of assessing the angularity of aggregates in the 1950s (Shergold, 1953b) and proposed a standard method based upon the amount of voids in a well compacted sample. This was subsequently adopted by British Standards Institution and appeared in BS 812 for many years. The value of such a test lies in the fact that the angularity of an aggregate affects the ease of handling of a mixture of the aggregate and binder (e.g. the workability of concrete) or the stability of mixtures that rely on the interlocking of the particles.

Apart from subjective descriptions derived from visual examination, earlier methods for individual particles were based on measurements of the radii of curvature of the particles or the relationship of the surface area to the volume. Although of academic value such methods are not suited for the routine testing of batches of aggregate. Consequently indirect methods were devised that include measuring the rate of flow of water through the sample (e.g. Carman, 1938), measuring the rate of fall of the aggregate through water (e.g. Schiel, 1941), measuring its behaviour on an inclined plane (e.g. Baturin, 1942), assessing the number of particles in a given volume or mass (e.g. Pickel and Rothfuchs, 1938), measuring the surface area (e.g. Director of Road Research, 1938), measuring the difference in percentages passing through square- and round-hole test sieves of the same nominal aperture (e.g. Schiel, 1948) and measuring the percentage voids in a sample when compacted in a standard way (e.g. Dunagan, 1940).

Three of these methods were explored in depth by Shergold: these were the measurement of the radii of particles, comparing percentages through square- and round-hole sieves and the determination of the voids. The latter was found to be the most satisfactory of the three and was found to correlate well with measurements of the compacting factor for concrete made with the aggregate. Although the standard method of test became known as the determination of "angularity number", the first form of the test expressed results in terms of "roundness index" (Director of Road Research, 1950). To determine this index the percentage voids contained in the aggregate was assessed and the roundness index was then derived using the formula:

Roundness index = 100 x (45 - per cent voids)/15.

Results ranged from 20 for angular crushed rock to 80 for a well-rounded beach gravel. They exhibited highly significant linear correlation with the compacting factor of concrete.

4.1.5.2 The determination of angularity number

In the determination of angularity number the aggregate was sieved to provide a sample sized between two adjacent test sieve sizes. It was then poured in three layers into a rigid metal cylinder of about 1/10 cu.ft. (0.003 cu.m.) capacity and subjected to tamping by 100 blows of a round-nosed steel tamping rod after pouring each layer. The cylinder was identical to that of similar capacity used in the bulk density determination. The sample was then weighed, the cylinder filled with water and weighed and the gross "apparent" relative density of the aggregate measured on a dry basis. The "angularity number" is then calculated from the formula:

$$\text{Angularity number} = 67 - (100\ M/CGa)$$

where

M is the mean mass of aggregate in the cylinder (g)
C is the mass of water required to fill the cylinder (g)
Ga is the relative density on an oven-dry basis.

Results ranged from 0 for a very rounded gravel to about 9 for a freshly crushed rock. This test appeared in BS 812 for several years: BS 812:1975 being the last revision in which it appeared.

4.1.5.3 Method of measuring degree of angularity

Methods of measuring angularity have also been studied by Lees at the University of Birmingham (Lees, 1961), where the shape of aggregates was regarded as having a great influence on many engineering, geological and geophysical properties. Those considered relevant to road engineering were:

i   The passage of grains through sieves.
ii The shear, compressive and flexural strengths of concrete, asphalt and macadam mixes.
iii The workability of concrete.
iv The adhesion of tar or bitumen to aggregate particles.
v   The skid resistance of road surfaces
vi The porosity, density and permeability of compacted or vibrated materials.

Lees studied the many methods of measuring aggregate sphericity that had been proposed and concluded that all were inadequate, particularly when measuring crushed particles. He also underlined the weakness that Shergold had already reported with regard to the BS 812 angularity test, namely that perfect spheres gave a higher angularity number (i.e. were more angular) than perfect cubes. He therefore developed a test for the determination of "degree of angularity" of particles which took account of:

i   The angle existing between considered faces, measured in the plane of the normals to these faces.
ii The number of angular corners.
iii The degree of projection of the tip of the corner from the centre of the main mass of the particle.

This test involved the individual measurement of all of the particles in the aggregate sample. Lees pointed out that although laborious, it was not more so than other direct measurements of sphericity and roundness. Because of the tedious nature of this test, it has not been adopted for use by either the engineer or the industry.

4.1.5.4 The loose bulk density method of measuring angularity

In view of the weaknesses of the method specified in BS 812, another alternative method was developed at the University of Birmingham. This was based on loose, rather than compacted, bulk density (Hughes and Bahramian, 1966). It was found to give satisfactory results with both spheres and natural aggregates and, in modified forms, could be used for both coarse and fine aggregates.

The results are expressed as "angularity factor" which is the ratio of the solid volume of uncompacted glass spheres in a standard container to the solid volume of the aggregate under test. Results ranged from 1.00 for glass marbles to 1.34 for "Aglite" (a synthetic aggregate). Two procedures were developed, one for fine and the other for coarse aggregates.

In the test for fine aggregate the sample is allowed to flow through a standard orifice into a 2 inch (50 mm) diameter cylinder until it overflows. The surface of the material is then struck off level with a straight edge and the solid volume of loose aggregate per unit volume (LV) is calculated from the formula:

$$LV = M/(PW)$$

where

> M is the mass of the aggregate
> P is the particle density of the aggregate
> W is the mass of unit volume of water

The angularity factor is then calculated by dividing this figure into the solid volume of glass beads of comparable size and under the same conditions.

For a coarse aggregate the method is similar except that the aggregate is inserted by means of a scoop and glass marbles used in the place of the beads.

4.1.5.5 Surface roughness

A method of assessing quantitatively the surface roughness of particles of concrete aggregate has been developed (Wright, P J F, 1955) to replace the visual classification of the surface texture of aggregates as "glassy", "smooth", "finely crystalline", "coarsely crystalline" or "porous". This method was devised to study the effect of the surface texture of aggregates on the strength of concrete. Surface roughness has also been considered to be of importance in determining the adhesion of binders to the surface and, in turn, the properties of the mix.

In this method particles of aggregates are embedded in a synthetic resin and thin sections prepared for microscopic examination using a modification of a technique developed at the Building Research Station for the examination of weathered building stones (Schaffer and Hirst, 1930). A tracing is then made of an enlarged image of the profile of the thin section of aggregate given by a projection microscope. The surface roughness of each aggregate particle is then assessed by measuring the length along the surface profile between two fixed points on the surface a short straight-line distance apart (about 2.5 mm), using a small map-measuring wheel. The result is expressed as a "roughness factor" which is the percentage difference between the profile length and the chord length between the two fixed points. The quantity ranges from about 2 for very smooth aggregates to about 16 for coarsely crystalline materials.

4.1.5.6 Flakiness and elongation

As clients frequently criticised the apparent flakiness and elongation of aggregates, and sometimes rejected them, an investigation was carried out (Shergold and Manning, 1953) to see whether such visual impressions were the same for different observers and how the results compared with the BS 812 flakiness index and elongation index tests.

Twenty eight observers rated seventeen samples of 1/2 inch (12.5 mm) chippings in order of merit. The agreement between observers was good. The test results identified the four worst samples, and most of the the four best, but there was more uncertainty about the intermediate samples. The results of the tests showed that the observers judged the chippings by flakiness rather than by elongation, and the number of flaky particles rather then their mass, and that they did not ignore the smaller particles in making their evaluation. It was concluded that the standard flakiness index test provided the engineer with a satisfactory way of specifying the shape characteristics of an aggregate.

Lees of the University of Birmingham has subsequently made a critical assessment of the various methods of assessing the flakiness and elongation of particles (Lees, 1964). He concluded that:

i   When measuring particles, the longest axis should be measured, together with an intermediate axis equal to the smallest circular (or square) aperture through which the particle can pass, and the narrowest axis as the width of the narrowest slot through which the particle can pass. (These three axes are not necessarily at right angles to one another.)

ii   The British Standard (BS 812:1960) flakiness and elongation gauges do not satisfactorily perform the function for which they were intended.

iii   Caliper measurements are the most accurate means of assessing flakiness ratio and elongation ratio, and give additional useful information about particle sphericity.

iv   Adjustable gauges such as the Rosslein gauge (Rosslein, 1941) were almost as accurate as calipers, and are quicker to operate.

v   Judgement by eye was more accurate than the British Standard gauges.

vi   A return to Zingg's ratio (Zingg, 1935) of 0.66 for both elongation and flatness ratio boundaries between shape categories was recommended. This compares with ratios of 0.6 and 0.56 respectively for the BS 812 flakiness index and elongation index tests.

Despite this advice these proposals have only been adopted for research purposes in Great Britain and have not displaced the British Standard gauges which are quick and easy to use and appear to be sufficient for specification and control purposes in the field of road engineering. However the consensus of international thought is in favour of either caliper tests or a more accurate version of the British Standard gauges.

4.1.5.7 The British Standard flakiness index test

For the purposes of BS 812 the "flakiness index" is defined as the percentage by mass of the particles in a sample of single-sized aggregate (that lying between two consecutive sieves in the series: 63.0 mm, 50.0 mm, 37.5 mm, 28.0 mm, 20.0 mm, 14.0 mm, 10.0 mm, 6.30 mm) whose least dimension (thickness) is less than 0.6 times the mean dimension of the two sieves. Gauges or slotted sieves of the appropriate slot width are used to separate the particles. The latest version of this test appears in BS 812:Part 105. A version is also under consideration for testing the flakiness of fine aggregates.

4.1.5.8 The British Standard elongation index test

For the purposes of this standard BS 812:1975 the "elongation index" is defined as the percentage by mass of the particles in a sample of single-sized aggregate (that lying between two consecutive sieves in the series: 50.0 mm, 37.5 mm, 28.0 mm, 20.0 mm, 14.0 mm, 10.0 mm, 6.30 mm) whose greatest dimension (length) is more than 1.8 times the mean dimension of the two sieves. Gauges with pins set with appropriated gaps are used to separate the particles by hand.

4.1.6 Polishing resistance

4.1.6.1 Polished-stone value determination

The design of the accelerated polishing machine in the 1950s (Maclean and Shergold, 1958) led to a satisfactory method of quantitatively assessing the merits of different aggregates. Refinement of the research apparatus led to the development of the present-day polished-stone value test of BS 812. The resistance to polishing of a sample of aggregate is determined by measuring the frictional characteristics (using a modified portable skid-resistance tester) of 40 to 60 selected 10-8 mm chippings mounted in a single flat layer in a suitable setting medium, which has been polished in a standard manner in an accelerated polishing machine. Corn emery and emery flour are used, respectively, to abrade and to polish their surfaces. This polishing, which is effected with a rubber tyre, is similar to that occurring in the road.

The result is expressed as "polished-stone value (PSV)", which is the effective coefficient of friction (x100) under the conditions of test. The test must be regarded as giving a relative rating for different aggregates rather than a measure of road friction. This is because the skid-resistance of roads is dependent on many factors other than the PSV of the aggregate (such as the temperature at the time of the test, the season of the year, the type of road site, the traffic intensity and the type of surfacing).

The PSV test has been specified in British Standard BS 812 since 1960. An account of its development and use is given in Chapter 6.

4.1.6.2 Polished-mortar value determination

This procedure (Franklin, 1978) is an adaptation of the method and apparatus for measuring the PSV. A sample of the fine aggregate is mixed with ordinary Portland cement to produce a mortar with an aggregate/cement ratio of 3.0 and a total-water/cement ratio of 0.6. Specimens of the same size as in the PSV determination are cast so that the upper, screed, surface is subjected to the polishing cycle. This is basically the same as for the PSV determination except for the omission of water during the first 3 hour period and minor adjustments to the nominal rate of feed of both grades of emery. Results are reported as "polished-mortar values (PMVs)".

4.1.6.3 Polished-paver value determination

The polished-paver value determination was developed by Lees at the University of Birmingham to provide a means of assessing the polish-resistance of clay pavers. The method employed makes use of the BS 812 aggregate abrasion method to polish the samples of pavers. The abrasion lap is modified by attaching a standard rubber disc, and corn emery and emery flour abrasives are fed to the samples under test in the same way as in the BS 812 PSV determination. The specimens are prepared to aggregate abrasion test specimen dimensions and flat control specimens (of the same control stone as the PSV test) are also made to these dimensions. After completion of the polishing procedure, the degree of polish is measured with the portable skid-resistance test in a similar way to the PSV test. A flat-to-curved correction factor is then applied to estimate the "polished-paver value", a correction being applied in the same way as the PSV test according to the level of the control specimens. The method has been published by the British Standards Institution as a draft for development (DD 155:1986).

4.1.7 Compactability tests

4.1.7.1 Modified vibrating hammer test

In this test (Pike and Acott, 1975) the depth constant for a standard mould is first determined by measuring the depth from the top of the empty mould to the anvil. A dry test specimen of 2.5 kg is well mixed and introduced into the mould in a way that minimises segregation, and then levelled. The anvil is placed on top of the specimen, the vibrating hammer located and surcharge load (360 N) applied. The sample is then compacted by operating the standard vibrating hammer for 180 seconds. The depth from the top of the mould to the anvil is then measured again.

The bulk density of the compacted aggregate is calculated by dividing the mass of the test specimen (in kilograms) by the volume of the specimen (in cubic metres). The "maximum proportion of volume occupied by solids" (MPVS) is derived by expressing the bulk density as a percentage of 10 x the relative density (oven-dry) of the aggregate. This is because the bulk density is expressed in kg/cu.m. and the relative density in Mg/cu.m.

4.1.7.2 Shear-box test

The machine used has a maximum shear-force capacity of 250 kN and operates at normal forces of up to 100 kN applied to specimens about 300 mm square in plan, with a continuously variable range of displacement rates between about 0.000000008 and 0.00008 m/s. The system employs a heavy-duty, vibrating hammer coupled to a double-acting hydraulic ram to accurately control the dry densities of test specimens. Compaction is effected through a foot which occupies the whole surface area of the mould.

4.1.8 Sieve Analysis

4.1.8.1 General

BS 812:Part 103 "Methods of determination of particle size distribution" specifies methods for carrying out particle-size analyses on all sizes of aggregate from a nominal 75 mm down to fine silt and clay sizes.

A study of the effect of sieve loading, particle size and duration of sieving (Shergold, 1946) on test results showed a close correlation between the percentage retained by the sieves used for fine aggregate and the sieve loading (See Figs. 9a and 9b). This loading effect was found to be determined by the proportion of "near-mesh" particles present and by their shape. The study indicated that these inaccuracies were more effectively remedied by reducing the sample size than by sieving for a longer period of time (see Fig. 10). Recommendations were given for the "maximum permissible amount of sand that should be retained" on each sieve size in order to avoid overloading. These provided the basis of our present specification for "minimum masses to be retained" on each sieve.

4.1.8.2 Particle size distribution

The particle size distribution of an aggregate is determined by shaking the sample in a prescribed manner through an appropriate succession of test sieves. The coarser sieves (5 mm aperture and above) are in perforated-plate for greater accuracy and the finer sieves are in woven-wire. Sieves normally used for roadmaking aggregates are specified in BS 410 and are as follows:

i   Perforated-plate sieves (all sizes in mm):
      75.0, 63.0, 50.0, 37.5, 28.0, 20.0, 14.0, 10.0, 6.30 & 5.00

ii   Wire cloth (in mm):
      3.35, 2.36, 1.70 & 1.18

iii   Wire cloth (in µm):
      850, 600, 425, 300, 212, 150 & 75 (sometimes 63)

Results are normally reported as the cumulative percentage by mass passing each appropriate test sieve and are, for many purposes, plotted on appropriate graph paper as a cumulative grading curve. For single-sized aggregates, however, it is more usual to report the percentage retained between successive sieves. Overloading of sieves during test can lead to serious errors and care should be taken to keep within the appropriate specified maximum sample masses to be retained on each sieve at the end of the sieving.

A specimen chart for recording sieve analysis results is included in BS 812:Part 103:1985. Information is also given on the preparation and cleaning of test sieves and their checking.

4.1.8.3 Determination of the particle size distribution of fine material

BS 812 specifies both sedimentation and decantation methods for the accurate determination of the particle size of fine material and a "field settling test" for use where a lower order of accuracy

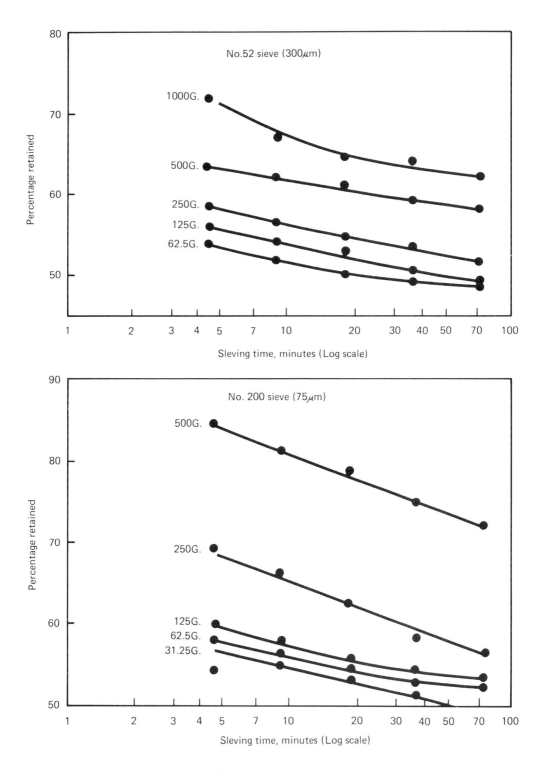

Fig. 9   Relations between sample weight, sieving time, sieve size and percentage retained.

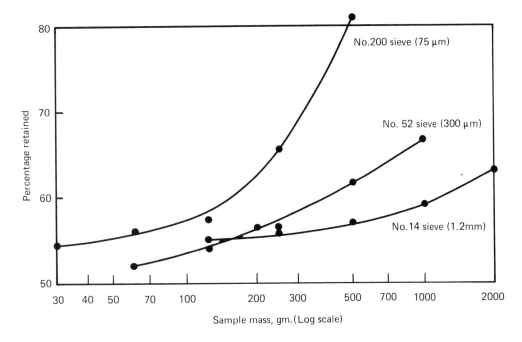

Fig. 10   Relation between sample weight and percentage retained.

can be tolerated. The sedimentation test (BS 812:Part 103.2:1989) estimates the percentage of material up to 20 μm in size by applying Stoke's law to the rate of fall of the particles from a suspension in water. The decantation test (BS 812:Part 103:1985) measures the percentage of all material passing a 75 μm BS test sieve using a washing and decantation technique. The field settling test gives a crude relative measure of the proportion of finer material in a sample of sand. The decantation method is preferred to dry sieve analysis if it is suspected that the presence of clay (or similar material) is hindering the separation on the 75 μm test sieve.

4.1.8.4 Precision of particle-size tests

The Transport and Road Research Laboratory, in conjunction with the Sand and Gravel Association, carried out a study of the errors in the sampling and testing of sub-base aggregates (Sherwood and Pike, 1984). The principal test studied was the determination of the particle-size distribution of these materials. The results are summarised in the Table 15.

The main findings of this study were:

i   The sampling error (that arising from the of taking bulk samples from the batch) was relatively small if the sampling was done according to the British Standard recommended procedure.

ii The within-laboratory error (r1) was relatively large. It was found that the error in sample reduction (that arising from the sub-division of the sample down to test portion) formed the major part.

TABLE 15
Summary of values for repeatability and reproducibility of the particle-size test carried out on Type 2 granular sub-base materials

Sieve size	Average % passing	Repeatability r1	Reproducibility R	R1	R2
20 mm	88	5	5	6	9
10 mm	75	7	6	9	12
5 mm	63	6	3	9	11
600 μm	36	4	3	5	7
300 μm	21	3	4	4	4
75 μm	9	1	2	2	3

iii   The between-laboratory error was quite small and, in consequence, the difference between repeatability and reproducibility was small. R, R1 and R2 being, respectively, the reproducibility of an identical sample, different test portions of the same laboratory sample and different laboratory samples (see Chapter 4.1).

iv   For sieve sizes where little material was retained the precision estimates were significantly lower and, for the others, precision was not greatly influenced by sieve size.

The findings of this study were surprising in two ways. Firstly they showed that sample reduction was a major source of error, whereas hitherto it had been regarded as being small, and secondly that the error in obtaining the initial sample was relatively small.

4.1.9   Tests for soundness

4.1.9.1 High temperatures

The effect of high temperatures that might be encountered by aggregates in a badly-designed or incorrectly-operated dryer has been investigated (Shergold, 1953c). It was found that to produce a significant reduction in the average strength of the eleven aggregates studied it was necessary to expose them for 24 hours at a temperature of 250 degrees C., 15 minutes at 500 degrees C. or 2 minutes at 1,000 degrees C. The extent to which the aggregates were affected differed considerably, but the results justified the conclusion that no general weakening of aggregate is likely to arise from the temperatures to which they are exposed in normal roadmaking practice.

4.1.9.2 Freezing and thawing

The effect on aggregates of repeated cycles of freezing and thawing has been studied (Shergold, 1954a) by subjecting fourteen different aggregates to 50 cycles of freezing at -25 degrees C. and thawing at 15 degrees C. The amount of disintegration that occurred was measured by sieve analysis and the samples were tested for resistance to crushing and abrasion before and after the freezing-and-thawing treatment. The main finding was that most rocks used in roadmaking are unlikely to be affected by frost. Fine-grained or glassy rocks and those with a water absorption greater than 1.6 per cent were considered to be those most liable to damage, and aggregates with water absorption values greater than 5 per cent were found to be highly susceptible to damage by frost.

This test procedure assessed the frost-susceptibility of roadstone samples and thus was different from the measurement of the frost-heave of materials such as sub-base aggregates.

4.1.9.3 Washington degradation test

The Washington degradation test was developed to differentiate between aggregates that would be suitable or otherwise for the construction of permanently stable dry stone bases (Minor, 1959). The test measures the amount of fine material passing a 75 μm test sieve produced by a sample of aggregate when it is agitated in a cylinder for 20 minutes. The fines are then subjected to a settling test and the "degradation factor" obtained from a measurement of the height of the sediment so produced. Aggregates giving values greater than 50 were found to be satisfactory whereas those giving lower values need to be used with caution. The test is used for igneous and metamorphic rocks only, the Texas ball mill test being used to assess sedimentary rocks (see below).

4.1.9.4 Slake durability test

In this test a number of aggregate particles are placed in a wire mesh drum, immersed in water, and rotated for 10 minutes. The loss in mass, expressed as a percentage, gives the slake durability index of the sample. The test is used for assessing the suitability of low-grade aggregates as sub-base materials.

4.1.9.5 Texas ball mill test

The Texas ball mill test (Texas Highways Department, Tex-116-3) is similar to the Washington degradation test, but is used for sedimentary rather than igneous rocks (see above).

4.1.9.6 Magnesium sulphate and sodium sulphate soundness tests

BS 812:Part 121 has been drafted to provide a standard method of determining unsoundness by the magnesium sulphate method. The principal method applies to 14.0 mm to 10.0 mm material. This is subjected to 5 cycles of immersion in a saturated solution of magnesium sulphate followed by oven-drying. This subjects the aggregate particles to the disruptive effects of the crystallization and repeated rehydration of magnesium sulphate within the pores of the aggregate. The result is expressed as the percentage of material finer than 10 mm that is produced. A subsidiary method is given for aggregate samples of different particle sizes. This draft British Standard also includes a method of determining the 10% fines durability factor. (See below).

An alternative form of this test uses sodium sulphate in the place of the magnesium salt. Versions of this test have been used by the Road Research Laboratory (e.g. Shergold and Hosking, 1963; Hosking and Tubey, 1969). Although critical of the sodium sulphate soundness test on the grounds of poor reproducibility and lengthy procedure, these workers concluded that it was the best test they had studied for detecting all types of unsound aggregates. They also suggested that the magnesium sulphate soundness test might be better. Both sodium and magnesium versions of the test have been standardised in the United States of America (ASTM C88).

4.1.9.7 Ten-per-cent fines durability factor

The determination of the 10% fines durability factor is given as an appendix to the draft BS 812:Part 121. The 10% fines test is applied to samples in both the dry and saturated surface dry condition. The factor is given by the difference between the two 10% fines values expressed as a percentage of the value obtained in the dry condition.

4.1.9.8 Methylene blue dye test

See 4.1.11 below.

4.1.9.9 Iron unsoundness

BS 1047:1974 specifies a test for iron unsoundness. A minimum of 12 particles of slag are immersed in water for 14 days. Slags which develop no "cracking, disintegration, shaling, checking or dusting" during this period shall be regarded as free from iron unsoundness.

4.1.9.10 Microscope test for blastfurnace slag

BS 1047 specifies a microscope test for the detection of "falling, dusting or lime unsoundness". Particles of between 5 mm and 2.36 mm in size or, alternatively, six particles of aggregate are embedded in a synthetic resin. The resulting specimen is ground and polished in a standard manner. It is then etched with a 10 per cent solution of magnesium sulphate for one minute at a temperature of 50 degrees C. The specimen is then examined microscopically for the presence of etched crystals. Their presence indicates that the slag is liable to "falling" etc.

4.1.10 Chemical tests

4.1.10.1 Acid-soluble material in fine aggregate

BS 812:Part 119:1985 specifies a method for the determination of acid-soluble material in fine aggregate. Carefully prepared test portions each of 50 g of material between 5.00 mm and 600 µm in size together with similar portions of the material smaller than 600 µm are treated with hot 4 mol/L hydrochloric acid until effervescence ceases (or up to 1 hour). The percentage of acid-soluble material is calculated and reported to the nearest 1 per cent. Results are reported for both size fractions separately. The test is intended to measure carbonate material and more prolonged digestion in the hot acid can make an undesirable addition to test results by a contribution from other acid-soluble components. This test is used to detect material that could have an adverse effect on skid-resistance. It is therefore only applicable to aggregates to be used in concrete and concrete paving blocks for road surfacing.

4.1.10.2 Water-soluble chloride salts

BS 812:Part 117:1988 specifies a method for the determination of water-soluble chloride salts. The test is essentially the same as that previously specified in BS 812:Part 4:1976. The sample is extracted with water to remove chloride ions and then analysed by the Volhard method. This method employs the addition of an excess of silver nitrate solution followed by back-titration with a standardized thiocyanate solution using ammonium iron (III) sulphate as an indicator. Results are reported as the chloride ion content expressed as a percentage.

4.1.10.3 Sulphate content

BS 812:Part 118:1988 specifies methods for the determination of the sulphate content of aggregates. It covers both the determination of water-soluble sulphate and also the total sulphate content.

Water-soluble sulphate ions are extracted with twice the sample mass of water and analysed either by an ion-exchange method or by a gravimetric method using barium chloride. The ion-exchange method has limitations and the gravimetric method is used where anions of strong acids are present and where sulphides are present (as in slags). This method is recommended for aggregates to be used for fill or hardcore. Results are expressed as the sulphate content of the 2:1 extract in grams/litre.

The assessment of total sulphate content is recommended for aggregates to be used in concrete. The sulphate is extracted with dilute hydrochloric acid, barium chloride is added and the precipitate of barium sulphate collected. The results are expressed either as the sulphate content of the 2:1 extract in grams/litre, or as the acid soluble sulphate as the percentage by mass of dry aggregate.

4.1.10.4 The chemical analysis of blastfurnace slag

BS 1047:1974 specifies methods for the determination of the total sulphur, acid soluble sulphate, silica, alumina, calcium oxide, magnesium oxide, iron oxide and titanium oxide contents of blastfurnace slag. The methods follow standard methods of chemical analysis. It is pointed out that although there are alternative methods of chemical analysis that might be used, greater uniformity of results would be achieved by using the BS 1047 methods and that, although rapid, their accuracy exceeds that required for highway engineering purposes.

4.1.11 Other tests

4.1.11.1 Cementation test

In the cementation test a mixture of coarsely crushed rock and water was ground to a stiff paste in a ball mill. Six briquettes were made from this paste in a mould under pressure, after which they were dried at room temperature for 20 hours and at 100 degrees to 110 degrees C. for 4 hours. Each briquette was then subjected to repeated blows in an impact machine, the height of fall of the hammer at each blow being 1 cm. The number of blows required to destroy the resilience of the specimen was taken as a measure of the "cementation value" of the rock.

This test was in considerable use at one time, but was falling into disuse by the time that British Standards were being first formulated at the turn of the century and so never achieved British Standard status. (Phemister *et al*, 1946).

4.1.11.2 The immersion wheel tracking test

This test was developed at the Road Research Laboratory (Director of Road Research, 1951; Lee and Nicholas, 1954; Mathews and Colwill, 1962).

The tracking machine consists of a wheel with a solid rubber tyre that is driven to and fro over a compacted sample of bituminous material immersed in water at a constant temperature of 40 degrees C. The machine operates at 25 cycles per minute and the test continues until either the sample fails or 48 hours have elapsed. The degree of penetration is recorded and a "break point" is reached when extensive stripping occurs, giving the failure time. Results are expressed as "failure time", this ranges from a few minutes to over 48 hours.

4.1.11.3 Drying shrinkage

BS 812:Part 120 1989 specifies two methods for testing and classifying drying shrinkage of aggregates in concrete. The methods are based on the earlier "Digest 35 method" (Building Research Station, 1963). The definitive method is a lengthy procedure requiring the preparation of sand/cement mortar prisms and measuring their length after 48 hour curing, this is followed by 26 day conditioning in a constant temperature/humidity oven and subsequent re-measurement at 7 day intervals until constant length is achieved. In the alternative method, oven drying at 105 degrees C. shortens the test. Nevertheless it is still a lengthy test requiring a pre-treatment of 5 days duration and a further 7 days of final immersion in water. Results are expressed as the percentage shrinkage: wet length less dry length, divided by dry length.

4.1.11.4 Methylene blue dye adsorption test

A study of the methylene blue dye adsorption test (Hosking and Pike, 1985) led to the development of two versions of the test, for small-size and large "real-size" samples respectively. The small-size method yielded higher values and better separation of the results: the real-size method showed less sampling error and saved time. The test is recommended as a screening test that would reduce the number of lengthy mortar prism tests (BS 812:Part 120) needed; however some limestones were found to give anomalous results (but failed safe). Versions of the test have been used to assess potentially unsound aggregates (Stewart and McCullough, 1985).

In the small-size test, portions are prepared so as to provide representative 1 g samples of material of between 1.18 mm and 600 µm in size. These are immersed in a standard dye solution for 24 hours after which the colour density is compared with a number of calibrated solutions. Results are expressed as "methylene blue adsorption value". The "real-size" version of the test is similar but uses test portions of actual-size aggregate of 1 kg in mass.

4.1.11.5 Shell content in coarse aggregate

BS 812:Part 106:1985 specifies a method for determination of the shell content of coarse aggregate. This procedure involves separating the sample into two size fractions (>10 mm and between 5 mm and 10 mm) and separating the shells and shell fragments by hand picking. Results are reported as the percentage of shell material in each size expressed to the nearest 1%.

4.1.11.6 Colour

The establishment of a scientifically based technique for assessing the colour quality of roadstones was developed by the Road Research Laboratory (Hosking and Ritson, 1968). The results of measurements with a Lovibond-Schofield tintometer were used to establish criteria for roadstones, for the three colour qualities of hue, saturation and luminance. This instrument conforms with Commission Internationale de l'Eclairage (CIE) requirements and is widely used for measuring colour.

A standard Lovibond cup and spinner were used to "average" the colours of a sample of aggregate. The sample consisted of washed stone of 12 mm to 9 mm in size, but the precise size was not considered to be of importance. The colour was assessed on wet stone; this was found to match the colour when polished by traffic. A series of colour categories were defined in terms of dominant hue wavelength, colour saturation and luminance factor.

The colour of stone in the road can, because of the "averaging" of the colour of the different crystals (and particles), appear different from that observed by close inspection of aggregate samples. Dust adhering to the surface of a stone can also greatly affect the apparent colour.

4.1.11.7 Frost heave

In this test (Roe and Webster, 1984) cylindrical specimens of aggregate are prepared by compacting the material at a pre-determined moisture content and density. These are placed in a self-refrigerated unit which has been designed to subject the upper surfaces of the specimens to freezing air at -17 degrees C., whilst their lower ends are allowed access to water at +4 degrees C. Water is drawn into the freezing zone where it forms ice and the specimen "heaves". The height of the specimens is measured at intervals over a period of 96 hours, and the maximum increase in height in mm gives the "frost heave". The test has been published by the British Standards Institution as BS 812: Part 124:1989.

4.1.11.8 California bearing ratio

The California bearing ratio (CBR) determination is specified in BS 1377 "Methods of test for soils for civil engineering purposes". This ratio is obtained from the relationship found between the force applied to and the penetration of a standard cylindrical plunger of cross-sectional area of 1935 sq.mm. into the sample under test. The CBR is the ratio of the force applied to a standard force at any value of penetration.

4.1.12 Classification

4.1.12.1 Trade classification

British Standard Specification BS 63 of August 1913 was the first of the many British Standard specifications to be concerned with roadmaking aggregates. This specification contained an appendix which set out twelve "trade names" of road-making rocks. These were drawn up by the Geological Survey and Museum to provide a relatively simple classification for road making purposes, as compared with the great variety of rocks that have been described and named by geologists and the considerable number of local and traditional names that were in general use. The method of classification was based on broad groups of similar rocks and the groups were named after a prominent rock type within the group (viz. granite, basalt, grit, gabbro, hornfels, limestone, porphyry, schist, flint, andesite, quartzite and artificial groups).

4.1.12.2 Knight's classification

Knight (Knight, 1935) considered that the physical tests of the time to be misleading when considered on their own, and that petrographical examination was essential in order to evaluate a roadstone. Roadstone tests have been greatly improved over the years but, for many purposes, petrological examination can usefully add to the assessment of a roadstone.

Knight was critical of the BS 63 classification of aggregates, and proposed what he considered to be a more scientific classification which was mainly based on their origin. Roadstones were first divided into artificial and natural categories. All artificial roadstones such as slags and clinker were included in the artificial group. The natural roadstones were sub-divided into igneous rocks, sedimentary rocks and the metamorphic group according to their mode of origin. Further subdivision of the igneous rocks led to the formation of the plutonic, hypabyssal and volcanic groups: similar division of the sedimentary rocks provided the calcareous, arenaceous and argillaceous groups. He thus reduced the number of groups to eight and provided groups that more accurately indicated aggregates of a similar nature.

Despite Knight's recommendations the British Standard grouping was retained without change until 1943 when, in BS 812:1943, a shorter list of eleven trade groups was adopted, and replaced that formerly in BS 63 (*viz.* artificial, basalt, flint, gabbro, granite, gritstone, hornfels, limestone, porphyry, quartzite, schist).

4.1.12.3 British Standard classification

The most recent British Standard comprehensive classification of aggregates (BS 812:Part 1:1975) was essentially the same as the "shorter" trade group classification described above. Table 16 (reproduced from this Standard) gives a list of the main rock types in each group.

The 1980s revisions of BS 812 do not include a comprehensive classification, but BS 812:Part 102:1984 "Methods of sampling" gives a list of the "rock types commonly used for aggregates" (see Table 17) for the purpose of petrological description. It also requires that the geological age be given for the sedimentary rocks, and states that "petrological description does not take account of suitability for any particular purpose, which should, be determined in accordance with the appropriate British Standard".

4.1.12.4 CADAM classification

From time to time alternative classifications have been proposed to replace that published by British Standards but, at the time of writing (1990), none has received wide general acceptance. The main contender has been the "classification and description of aggregate materials" (CADAM) classification. This is based on "form", "class" and supplementary information.

TABLE 16
Main rock types (BS 812:1975)

1. Artificial group
crushed brick
slags
calcined bauxite
synthetic aggregates

2. Basalt group
andesite
basalt
basic porphyrite
diabase
dolerites of all kinds
(including theralite and
teschenite)
epidiorite
lamprophyre
quartz-dolerite
spilite

3. Flint group
chert
flint

4. Gabbro group
basic diorite
basic gneiss
gabbro
hornblende-rock
norite
peridotite
picrite
serpentine

5. Granite group
gneiss
granite
granodiorite
granulite
pegmatite
quartz-diorite
syenite

6. Gritstone group
(including fragmental
volcanic rocks)
arkose
greywacke
grit
sandstone
tuff

7. Hornfels group
contact-altered rocks of
all kinds except marble

8. Limestone group
dolomite
limestone
marble

9. Porphyry group
aplite
dacite
felsite
granophyre
keratophyre
microgranite
porphyry
quartz-porphyrite
rhyolite
trachyte

10. Quartzite group
ganister
quartzitic sandstones
recrystallized quartzite

11. Schist group
phyllite
schist
slate
all severely sheared rocks

TABLE 17
Rock types commonly used for aggregates (BS 812:1984)

Petrological term	Description
andesite*	a fine grained, usually volcanic, variety of diorite
arkose	a type of sandstone or gritstone containing over 25% feldspar
basalt	a fine grained basic rock, similar in composition to gabbro, usually volcanic
breccia**	a rock consisting of angular, unworn rock fragments, bonded by natural cement
chalk	a very fine grained Cretaceous limestone, usually white
chert	cryptocrystalline*** silica
conglomerate**	a rock consisting of rounded pebbles bonded by natural cement
diorite	an intermediate plutonic rock, consisting mainly of plagioclase, with hornblende, augite or biotite
dolerite	a basic rock, with grain size intermediate between that of gabbro and basalt
dolomite	a rock or mineral composed of calcium magnesium carbonate
flint	cryptocrystalline*** silica originating as nodules or layers in chalk
gabbro	a coarse grained, basic, plutonic rock, consisting essentially of calcic plagioclase and pyroxene, sometimes with olivine
gneiss	a banded rock, produced by intense metamorphic conditions
granite	an acidic, plutonic rock, consisting essentially of alkali feldspars and quartz
granulite	a metamorphic rock with granular texture and no preferred orientation of the minerals
greywacke	an impure type of sandstone, composed of poorly sorted fragments of quartz, other minerals and rock; the larger grains are usually strongly cemented in a fine matrix
gritstone	a sandstone, with coarse and usually angular grains
hornfels	a thermally metamorphosed rock containing substantial amounts of rock-forming silicate minerals
limestone	a sedimentary rock, consisting predominantly of calcium carbonate
marble	a metamorphosed limestone
microgranite*	an acidic rock with grain size intermediate between that of granite and rhyolite
quartzite	a metamorphic rock or sedimentary rock, composed almost entirely of quartz grains
rhyolite*	a fine grained or glassy acidic rock, usually volcanic
sandstone	a sedimentary rock, composed of sand grains naturally cemented together
schist	a metamorphic rock in which the minerals are arranged in nearly parallel bands or layers. Platy or elongated minerals such as mica or hornblende cause fissility in the rock which distinguishes it from gneiss
slate	a rock derived from argillaceous sediments or volcanic ash by metamorphism, characterized by cleavage planes independent of the original stratification
syenite	an intermediate plutonic rock, consisting mainly of alkali feldspar with plagioclase, hornblende, biotite, or augite
trachyte*	a fine grained, usually volcanic, variety of syenite
tuff	consolidated volcanic ash

* The terms microgranite, rhyolite, andesite, or trachyte, as appropriate, are preferred for rocks alternatively described as porphyry or felsite.
** Some terms refer to structure or texture only, e.g. breccia or conglomerate, and these terms cannot be used alone to provide a full description.
*** Composed of crystals so fine that they can be resolved only with the aid of a high power microscope.

i   Form. The form of the aggregate is whether it is crushed rock, gravel, etc.

ii   Class. There are four classes: "carbonate", "quartz", "silicate" and "miscellaneous". These are based on the predominant mineral (or group of minerals). The silicate class is further subdivided into "igneous", "metamorphic" and "sedimentary" sub-groups.

iii   Supplementary information. This consists of geological age for sedimentary rocks, colour and grain size for igneous and metamorphic rocks, and foliation (or fissibility) for both metamorphic and sedimentary rocks.

A comprehensive account of the system is given in the Engineering Geology Special Publication No.1 (Geological Society, 1985).

4.2 Correlation between tests

4.2.1 Petrology and other characteristics

4.2.1.1 General

From early times there have been numerous attempts to relate the petrological characteristics of roadstones and aggregates to their performance in the road and to the results of measurements of other characteristics. The petrological characteristics have been assessed in a number of different ways ranging from the classification of roadstones and aggregates into a number of general petrographical types, such as granites and limestones, to the detailed examination of characteristics such as mineral grain size and proportions of the different minerals.

Generally, results have been disappointing, but there have been numerous good correlations found between some physical and mechanical properties and petrographical characteristics for a number of particular rock types. However these relationships have usually failed when applied to other rock types.

4.2.1.2 Work of Lovegrove, Flett and Howe

A summary of the general conclusions drawn by Lovegrove and the Geological Survey and Museum (Lovegrove, *et al*, 1929) regarding correlation between the physical and mechanical properties of aggregates and their petrography is given below. These conclusions are as true today as they were at the time of their publication.

    a.   *Granite*

The presence of finely crystalline material to serve as a binding or cementing substance is important for good resistance to attrition. Microscopic cracking in quartz is related to poor behaviour under the attrition test. The presence of large feldspar crystals impairs the toughness of granite.

b.   Diorite, epidiorite and diabase

If much hornblende is present in these rocks, they resist attrition and are tough.

c.   Porphyry

The average resistance of the rocks of the porphyry group to attrition, abrasion, impact and crushing is high and is related to the fine grain, the hard nature and the intergrowth of the constituent minerals. The explanation of the lower resistance offered by some rocks of this group is not always clear but seems to depend on the weathered condition of the stone.

d.   Basalt and dolerite

Rocks of these types vary greatly in the state of preservation and there is corresponding variation in their behaviour under test. Finely crystalline rocks are tougher than the coarse-grained. Rocks with fresh olivine are harder, more resistant to attrition and tougher than quartz-dolerites, basalts and dolerites in which the olivine is serpentinised.

e.   Limestone

Limestones of Jurassic age are not so firmly compacted and cemented as those of Carboniferous age. The latter are uniform in their resistance to attrition whatever the relative proportion of organic debris and crystalline calcite, but the toughest appear to be those which are least crystalline and which consist of small broken fragments of fossils of all shapes and sizes cemented by a small quantity of secondary calcite. Dolomites of Carboniferous age have superior resistance to attrition to the corresponding limestones, but are apparently neither harder nor tougher.

f.   Sandstone

Among rocks of the same type, the geologically older are more resistant to attrition. The behaviour of the stone depends on the nature of the cementing material. Where the cement is quartz, toughness and resistance to attrition are great and such sandstones grade into quartzite. When the cement is mixed micaceous and silty material, or calcareous, or when the rock is very porous and weak, toughness and resistance to attrition suffer.

g.   Flint

Flint is very resistant to attrition but is brittle.

4.2.1.3 A study of quartz-dolerite roadstones

By the mid-1950s it had become generally recognised that variation in the physical and mechanical properties of rocks could be correlated only very broadly with petrographic characteristics. An investigation (Sabine *et al*, 1954) was made on a group of rocks of limited petrographic type in order to ascertain whether any relation between mineral content and test results could be observed. The rock type chosen was quartz-dolerite.

Fifty-five samples of British quartz-dolerites were examined petrographically and by a number of physical and mechanical tests. These included tests for aggregate crushing value, aggregate impact value, aggregate abrasion value, water absorption and apparent specific gravity (particle density). Other tests were for Los Angeles abrasion value, dry attrition value, wet attrition value, impact value and crushing strength.

Correlation was attempted by plotting the test values against the percentage of each mineral present (107 plots in all). Additionally correlation was sought for different groups of rocks, such as of different grain size and freshness. In general no correlation was found. The best, between iron-ore content and particle density was still described as "poor". No correlation was found between grain size determinations and the results of the tests. The only satisfactory correlation found was between the geological occurrence and the strength and water absorption of the quartz-dolerites. Samples from the Whin Sill being generally best, followed by the Tertiary dykes, with the Permo-Carboniferous samples giving the poorest results.

The authors concluded this investigation by recording that:

> "In practice it is not sufficient to show that a group of rocks as a whole has certain petrographic characters and that these can be correlated with the results of physical testing or with service behaviour; what is required is to be able to establish with reasonable certainty from the characters of any member of the group what its behaviour on physical testing or under service conditions will be. The present investigation has not shown that this can be done."

4.2.1.4 Petrography and polishing

The Geological Survey and Museum have made a study of the petrological aspects of the polishing of natural roadstone (Knill, 1960), in which the petrological characteristics of members of the different aggregates groups were compared with their resistance to polishing. A summary of her results was:

i   The gritstone group was outstandingly good, with the resistance to polishing being always high whereas the limestone and flint groups yielded the lowest resistance. The other groups studied, basalt, granite and quartzite, yielded intermediate results.

ii   The resistance to polishing of the 46 samples of the basalt group showed a wide range. Resistance was higher when minerals of different hardness were present, and when the ground-mass was foliated or fluxioned. The resistance was also influenced by the proportion and hardness of secondary minerals, softer minerals giving the higher resistance.

iii   In the groups of igneous rocks the petrological characteristics which most readily affected the resistance to polishing were variation in hardness between the minerals and the proportion of soft minerals present. Rocks with cracks and fractured minerals were of higher resistance, whereas finer-grained allotriomorphic rocks tended to polish more readily.

TABLE 18
Correlation between polished-stone coefficient and the petrography of specimens tested in the limestone group

PSC	Average grain size (mm)	Organisms	Coarse crystalline carbonate patches	Insoluble residue (%)
0.30	0.02	sparse	+	1.90
0.35	0.25	-	+	1.42
0.35	0.02	sparse	+	1.40
0.35	0.02	sparse	+	2.30
0.35	0.26-0.28	sparse	-	-
0.40	0.1	sparse	+	9.80
0.40	0.035	-	+	2.80
0.40	0.19-0.28	-	-	4.60
0.40	0.02-0.03	frequent	+	-
0.40	0.08-0.10	-	-	3.10
0.45	0.02-0.03	sparse	+	1.26
0.45	0.09	-	-	2.30
0.50	0.02-0.03	sparse	+	5.00
0.55	0.02	plentiful	-	3.70
0.55	0.02	plentiful	+	36.0

iv   The gritstones gave a uniformly high resistance to polishing since hard crystals and lithic fragments pluck out of a generally soft and friable matrix.

v   Higher resistance in the limestone group appeared to be related to the presence of insoluble residue (especially if quartz) and to the presence of coarsely crystalline patches of carbonate. Table 18 shows the correlation between petrography and polished-stone coefficient (PSC) for this group.

4.2.1.5 Gritstone survey

The Institute of Geological Sciences and the Transport and Road Research Laboratory carried out a survey of British resources of arenaceous rocks (gritstones and similar rocks) with the object of finding new sources of aggregate for skid-resistant surfacings (Hawkes and Hosking, 1972). The survey disclosed large resources of high-quality material and has provided information that will be of value in future searches for similar material and in selecting stone within a quarry. Results showed that desirable qualities (high resistance to polishing and abrasion) in arenaceous rocks are more dependent on their geological history than on their mineral composition. The following relationships were found between the petrology and physical characteristics of these rocks:

i   Polished-stone value (PSV), aggregate abrasion value (AAV) and aggregate impact value (AIV) were all found to correlate significantly with both ultra-sonic velocity and effective porosity (measures of the degree of consolidation of a rock).

ii   PSV, AAV and AIV were also found to correlate significantly with the proportions of quartz fines, quartz fragments and, to a lesser extent, clay, feldspar and lithic fragments.

iii Grain density was also found to be affected significantly by the proportions of the above constituent minerals.

iv Multiple correlation was found between polished-stone value, aggregate abrasion value and aggregate impact value (See Fig. 11).

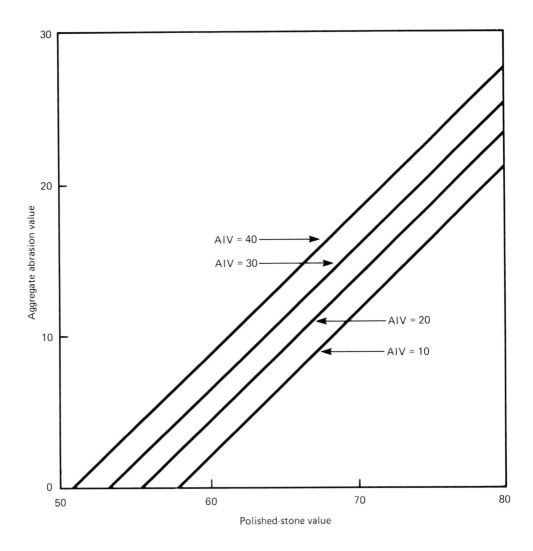

Fig. 11   *Multiple correlation between AAV, PSV and AIV.*

4.2.1.6   Petrography and polished-stone value

a.   Igneous rocks

A petrographical comparison has been made of the two control stones that have been used for the PSV determination of BS 812 (West and Sibbick, 1988b). Although the original control stone was an alkali granite and the new one is a dolerite, they both give a similar PSV. The study showed that for rocks of this type (crystalline igneous rocks) the polish imparted to a roadstone specimen in the accelerated polishing test depends on the overall hardness of the rock. The petrographic examination showed that the two rocks had virtually the same mean hardness (761 and 720 Vickers hardness number respectively). The authors point out that other textural features would also affect the polishing characteristics of sedimentary and metamorphic rocks.

b.   Limestones

A study of the polishing characteristics of limestones (Shupe and Loundsbury, 1959) showed good correlation between the resistance to polishing of twelve limestones and their calcium carbonate content. Resistance to polishing was expressed as "relative resistance value" as determined in a wear and polishing procedure developed at Purdue University. Their studies had already shown lack of consistency of structure or of grain hardness appeared to be essential for good skidding resistance. The behaviour of the limestones was in keeping with this observation as the more nearly a limestone approaches to pure calcium carbonate, the more uniform would be the structure and hardness, so that the resistance to polishing would be low. Fig. 12 shows similar findings by the Author (Hosking, 1970).

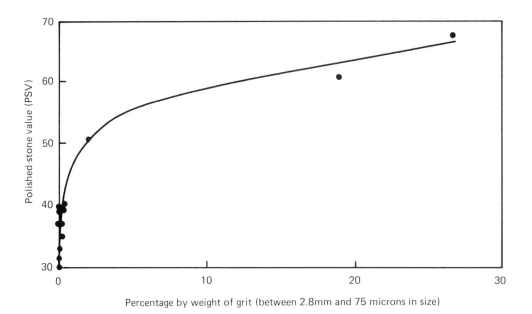

Fig. 12   The relationship between the polished-stone value and percentage grit in samples from 19 limestone quarries.

4.2.2 Correlation between mechanical tests

4.2.2.1 Lovegrove attrition test and Deval attrition test

A series of tests in wet and dry attrition were made at the NPL (National Physical Laboratory, 1913/14)) in order to obtain the relation between the results given by the Deval attrition test and the Lovegrove attrition test. The relationships obtained were:

Coefficient of wear = 40/% loss in Deval machine
Coefficient of wear for wet tests = 155/% loss in Lovegrove's machine
Coefficient of wear for dry tests = 75/ % loss in Lovegrove's machine

Lovegrove reported that although there was general agreement between the two tests, considerable discrepancies were frequently observed. These were attributed to the additional impact action of the Lovegrove attrition test, whereas the Deval attrition test gave a simply rotary action.

4.2.2.2 Aggregate crushing test and Los Angeles abrasion test

Studies were made of the correlation between crushing strength, Los Angeles abrasion value, aggregate crushing value, dry attrition value, wet attrition value, abrasion value and impact value (Phemister *et al*, 1946). These showed almost numerical equality for the aggregate crushing and the Los Angeles tests, and poorly defined relationships between the other tests.

During the development of the aggregate crushing test comparisons of results were made between the aggregate crushing values and Los Angeles abrasion values for a range of roadstones (Markwick and Shergold, 1945). The comparisons (Fig. 13a) showed close numerical agreement between the results of the two tests. This good agreement meant that the preferred test (the aggregate crushing test) could take advantage of the good correlations that had already been established between Los Angeles abrasion values and performance. The correlation obtained for aggregates over a wider range of strength is shown in Fig.13b; this shows that the linear relationship fails for aggregates with Los Angeles abrasion values greater than about 30 and demonstrates the less sensitive nature of the aggregate crushing test when applied to weaker aggregates.

4.2.2.3 Aggregate crushing test and aggregate impact test

Fig. 14 shows the close numerical agreement between these tests that was found during the development of the aggregate impact test.

4.2.2.4 Tensile and crushing strength

A study was made of the tensile strength of a range of roadmaking rocks (Hosking, 1955), and the results were compared with those obtained for their crushing strength.

The indirect tensile strength of a rock was found to be about one-twentieth of the equivalent crushing strength. Flexural strengths showed more variation but were normally about one-tenth of the crushing strength. The ratio of flexural strength to crushing strength varied from about 1:7 to 1:25 and it was generally higher for the more homogeneous rocks and lower for the non-homogeneous and coarse-grain rocks.

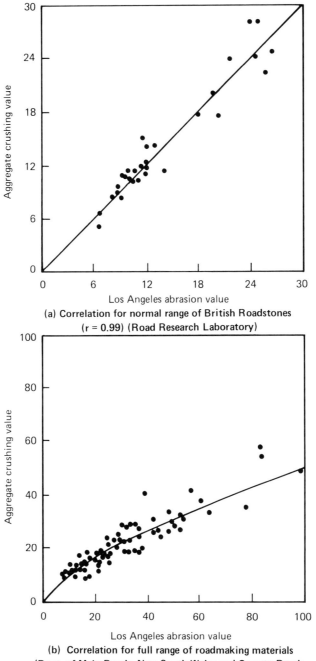

Fig. 13   Correlation between Los Angeles abrasion test and aggregate crushing test.

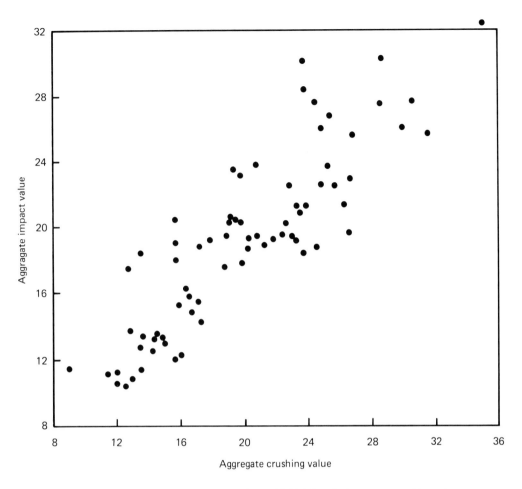

Fig. 14   Correlation between aggregate impact value and British Standard aggregate crushing value.

4.2.2.5 Angularity number and compacting factor of concrete

It has been shown (Shergold, 1953b) that there was good linear correlation between the determination of angularity number and the compacting factor for concrete. Fig.15 shows the correlation obtained with seven different aggregates which were used to make concrete with a standard aggregate grading, cement-aggregate ratio, water-cement ratio and the same cement and fine aggregate (Compare with Fig. 4).

4.2.2.6 Angularity number and MPVS

SAGA's research team at TRRL carried out an investigation of the mechanical properties of graded aggregates. In the course of this work (Pike, 1972) they made a comparison of the results of angularity number tests (BS 812:1967) with the "maximum proportion of volume occupied by solids" (MPVS) obtained with the British Standard vibrating hammer compaction test

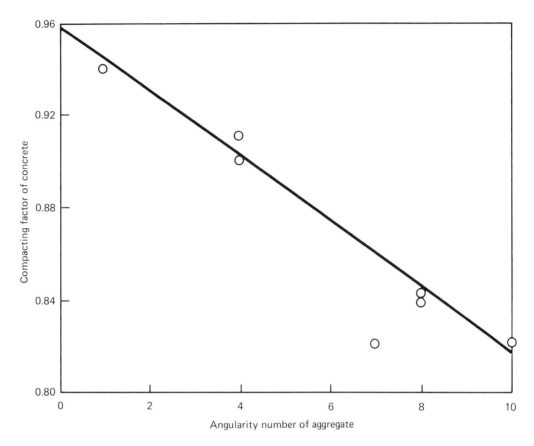

Fig. 15   Relationship between the angularity number of aggregate and the compacting factor of concrete made with it.

(BS 1377:1967). Comparative tests were made on seventeen "1 1/2 inch down" aggregates of a standard grading having a coefficient of uniformity of 47. The angularity number was measured for the 19-13 mm, 13-9.5 mm and 9.5-6.3 mm sizes, and averaged.

It was pointed out that the angularity number cannot be used to assess the compactability of graded aggregates; also that the type of compaction used in this test is neither similar in nature nor in degree to that used in the field. Nevertheless, a highly significant correlation (99 per cent probability level) was obtained between the values (see Fig. 16). This shows that, provided the grading is kept constant, an increase in angularity and roughness of the coarse particles produces a reduction in compactability.

4.2.2.7 Coefficient of uniformity and MPVS

The coefficient of uniformity is defined as the ratio of that aperture size through which 60 per cent of the aggregate passes, to that aperture through which 10 per cent passes. Well-graded size distributions give high values (> about 40) and poorly-graded and uniformly-graded distributions give low values.

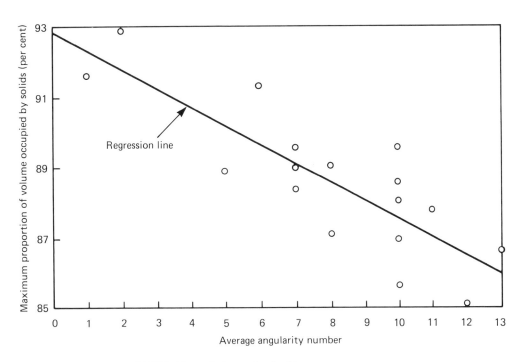

Fig. 16   Correlation of MPVS with angularity number for 17 aggregates.

Work by SAGA (Pike, 1972) included a comparison of the coefficient of uniformity of two aggregates, each with a range of gradings, with the MPVS obtained with the British Standard vibrating hammer compaction test (BS 1377:1967). The aggregates were a smooth rounded quartzite gravel and an angular and rough crushed gritstone. The gradings were chosen to represent the broad range of size distributions in common use in sub-base, roadbase and base-course materials. They included well-graded, poorly-graded and gap-graded distributions.

In Fig. 17 values of MPVS are plotted for the two aggregates against the logarithms of the coefficients of uniformity derived from wet-sieve analyses made on samples after compaction. Highly significant correlations (at the 99 per cent probability level) were obtained for both aggregates.

4.2.2.8 Aggregate crushing value and crushing strength

During the development of the aggregate crushing test comparisons of results were made between the aggregate crushing values and crushing strengths of range of roadstones (Markwick and Shergold, 1945). The results, Fig. 18, show rather poor linear correlation. The scatter is caused by two factors. The first is the considerable variability of the results of crushing strength tests, which are carried out on individual specimens. The second is a result of the aggregate crushing test results being affected by properties such as particle shape as well as crushing strength.

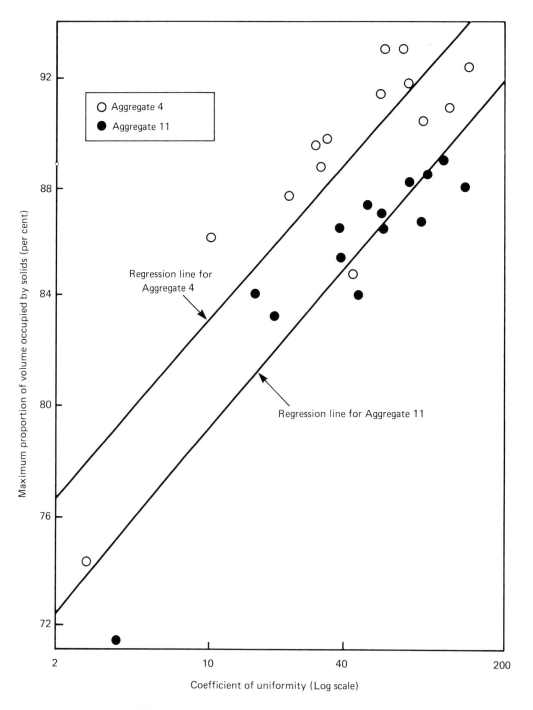

Fig. 17   Correlation of MPVS with coefficient of uniformity for two aggregates.

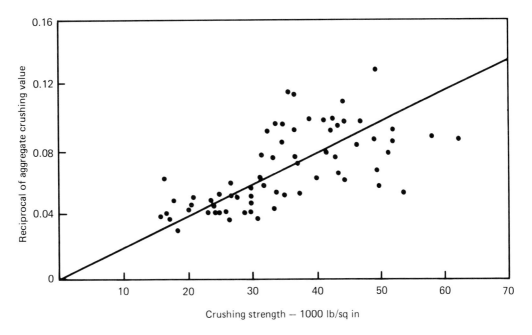

Fig. 18   Correlation of aggregate crushing value and crushing strength.

4.2.2.9 Dry attrition value and crushing strength

Fig. 19 shows the correlation that was obtained when the results of 400 tests for crushing strength were compared with the reciprocal of the dry attrition values of the same roadstones (Shergold, 1948). It was considered that the relatively poor correlation (correlation coefficient 0.62) was due partly to the poor reproducibility of the attrition test as well as the difference in the properties being measured in the two tests.

4.2.2.10 Polished-stone value and other properties

The relation between PSV and skidding measurements is discussed in Chapter 5. No general correlation between PSV and other test results has been found, but, sometimes, within particular groups of aggregates, a closer relationship exists. Results obtained by Hopkins in the early days of the polishing test, showed no significant correlation between polished stone coefficient (PSC) (PSV = approximately 90 x PSC) and aggregate abrasion value (AAV) in the limestone group. For igneous rocks an increase in PSC was accompanied by a significant decrease in abrasion resistance. The comparable relationship for gritstones was similar, but changes in PSC were accompanied by much greater changes in AAV than was found with the igneous rocks (Hopkins, 1959).

Fig. 20 shows a comparison of PSV and AAV for a much wider range of gritstones (Hosking, 1970). A similar comparison for blastfurnace slags is shown in Fig. 21 (Hosking, 1970).

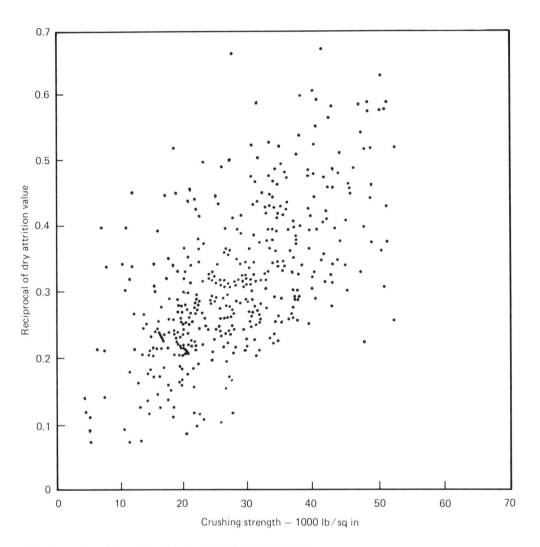

Fig. 19   Correlation of dry attrition value and crushing strength.

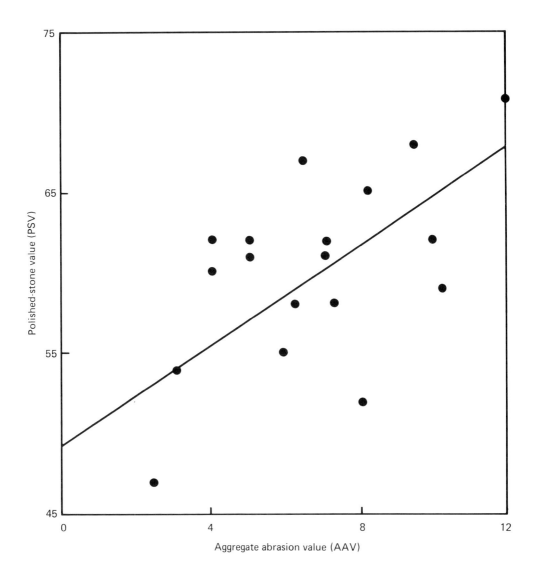

Fig. 20 The relation between PSV and aggregate abrasion value for 19 gritstones.

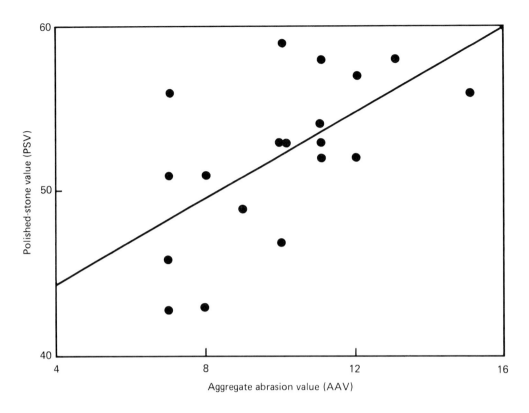

Fig. 21 The relation between PSV and aggregate abrasion value for 20 blastfurnace slags.

The effect of changes in bulk density on the PSVs of blastfurnace slags is shown in Fig. 22 (Hosking, 1970).

4.2.2.11 Methylene blue dye test and the Digest 35 shrinkage test

Fig. 23 shows a correlation of the blue dye test with the Digest 35 shrinkage test (Hosking and Pike, 1985).

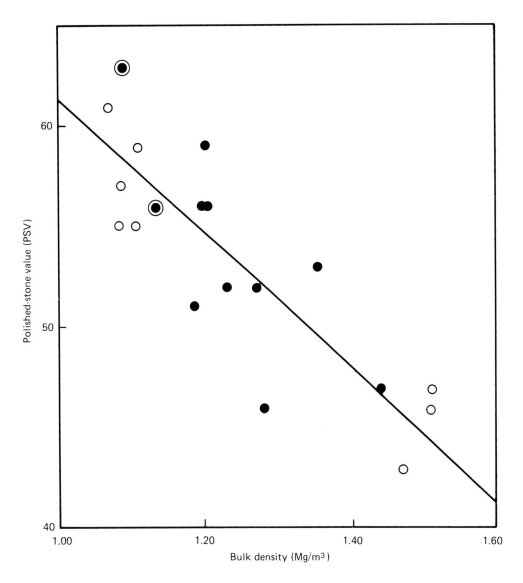

Fig. 22   *The relation between PSV and bulk density of some blastfurnace slags.*

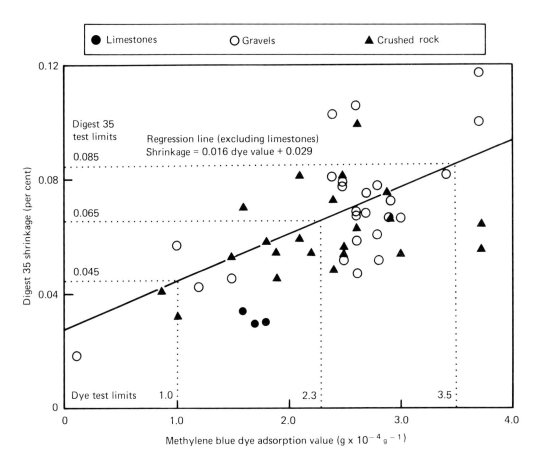

Fig. 23   Relationship between "Digest 35" shrinkage and the results of methylene blue dye adsorption test.

5 Special requirements of aggregates for skid-resistance

5.1 Bituminous surfacings

5.1.1 Development of the polished-stone value test

5.1.1.1 Background

After the 1939-1945 war the large increase in motor vehicles and higher traffic speeds gave additional impetus to the need to study the relation between road materials and the skid-resistance of roads. In 1952 an accelerated wear machine was constructed by the Road Research Laboratory to study the different factors that cause wear on a surface dressing. Provisional results suggested that loose grit on the surface might be an important factor contributing to the abrasive wear of aggregates, and that the machine showed promise as a method of studying the polishing action of traffic.

At the same time a simple machine was devised by the Road Research Laboratory to study the degree to which various types of aggregates preserved their sharp edges during their life on the road. In this machine one of the edges of a chipping was subjected to 600 applications of stress per minute by forcing it, with a thrust of 20 lb. force (9 kg force), against the tread of a pneumatic tyre which caused the latter to rotate slowly. In this way the equivalent to a year's wear on a heavily trafficked road was simulated in a few hours.

The profile of the chipping was photographed before and after the test, and the final profile was obtained after enlargement which had a linear magnification of 100. The "shape-change factor" was then assessed for measurements of chord lengths along the profile. The results showed that fine-grained limestones became smoother and coarse-grained dolerites remained rough.

Having established that the shape of the projecting edges of a chipping can change with wear, an apparatus was then devised to measure the effect of these changes in terms of coefficient of friction between the chippings and tyre-tread rubber. The specimen under test in this apparatus was bonded to an anvil held in a carriage running on guides. The specimen was then held under a load against a rubber track, and the carriage driven along at the desired speed of test by compressed air. As the slide travelled over the rubber, the frictional force was recorded, together with signals from which the speed and position of the slider at any instant could be deduced.

By this time the research with the accelerated wear machine had shown promise as a research tool for investigating the polishing of aggregate by traffic and had the advantage of testing a specimen consisting of 50 chippings instead of just one. The single-chipping apparatus was therefore confined to basic studies of the effect of the shape of projections on their wet skid-resistance (See Chapter 6.1.2).

Meanwhile the accelerated wear machine was used to test a surface formed by 1/2inch (12.5 mm) chippings of 14 different types of aggregate, set in sand-cement mortar round the rim of a wheel, each aggregate type covering an area of approximately 6 sq.ins. (3900 sq. mm.). In most other respects it was similar to the present-day accelerated polishing machine used in the polished-stone value (PSV) determination. However a braking force could be applied to the tyre by means of a hydraulic mechanism, to produce slip between the tyre and the "road". The road was kept wet by spraying with water. (During "dry" tests the tyre over-heated and burst.) When aggregates were subjected to rolling action alone in this machine little polishing occurred, but polishing did occur when fine grit was fed between the aggregates and the tyre. The original grit was detritus collected from under the wings of cars, but later work employed prepared gradings of sand.

Studies with this apparatus showed that many roadstones polished under the abrasive action of tyres and detritus, and that this polish made the road surface slippery. It was found that the final degree of polish attained depended not only on the properties of the roadstone, but also on the fineness of the sand, and that the polish could be removed by substituting coarser grit between the tyre and the roadstone.

The rate at which the polishing occurred was found to depend on the properties of the roadstone - some polished more slowly than others but still attained a fairly high degree of polish. It was also found that the rate could be accelerated by introducing slip between the tyre and the roadstone.

5.1.1.2 Roughness number test

By 1957 the accelerated wear machine was being used as a basis of a test for assessing the resistance to polishing of aggregates. The extent of the polish was measured by means of the portable skid-resistance tester and was expressed as a "roughness number" which ranged from 0 for specimens that become highly polished to 10 for specimens which remained rough after the test.

The significance of the results of the accelerated polishing test was then assessed in two ways:

i   A range of roadstones were used as the aggregates in a number of the Laboratory's full-scale experimental surfacings which carried heavy traffic, and their roughness numbers compared with the average sideway-force coefficient (SFC) at 30 mile/h (50 km/h) of the surfacings when wet. After the roads had been carrying traffic for 18 months it was found that the fluctuations in SFC were purely of a seasonal character.

ii   Small specimens of surfacing each consisting of a layer of chippings embedded in mortar were inserted flush with the surface of a road carrying heavy traffic, and the "roughness numbers" of these areas of chippings were determined in situ after various periods of time and compared with the roughness numbers of the chippings when subjected to the accelerated polishing test. (See also Chapter 5.5.2 "Small-scale road experiments".)

The correlation in both cases was satisfactory, showing that the method of polishing employed in the test was acceptable.

In the course of tests on the areas of chippings, fluctuations in the degree of polish were observed.

These fluctuations were found to be associated with variations in temperature at the time of testing and with the wetness of the road surface: dry weather was associated with an increased polishing while a prolonged period of wet road conditions resulted in decreased polishing. These effects were thought to have a bearing on the seasonal variation in the skidding resistance of road surfacings.

5.1.1.3 Polished-stone coefficient test

In 1958 the "polished-stone coefficient" (PSC) replaced "roughness number" as the measure of polish resistance. Good correlation was found between PSC and the skid-resistance value of road specimens, measured with the portable tester. Good correlation was also found between PSC and the SFC of surface dressings, coated macadams and asphalts in which the stone formed the greater part of the wearing surface. This correlation was found to hold even where the traffic on the roads concerned was as low as 5,000 tonnes per day. Examples of these early correlations are shown in Figs. 24 and 25.

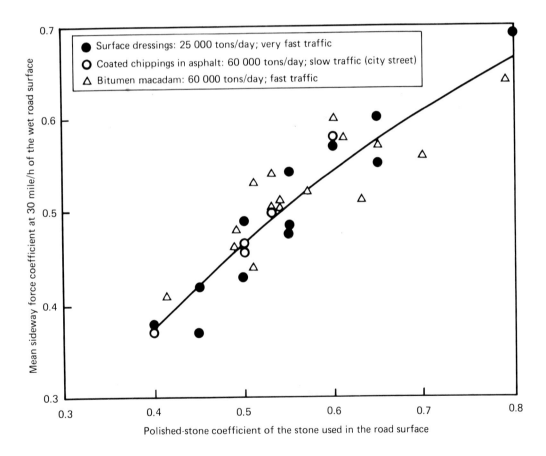

Fig. 24 Comparison of the mean sideway-force coefficient at 30 mile/h of the wet road surface with the polished-stone coefficient of the stone.

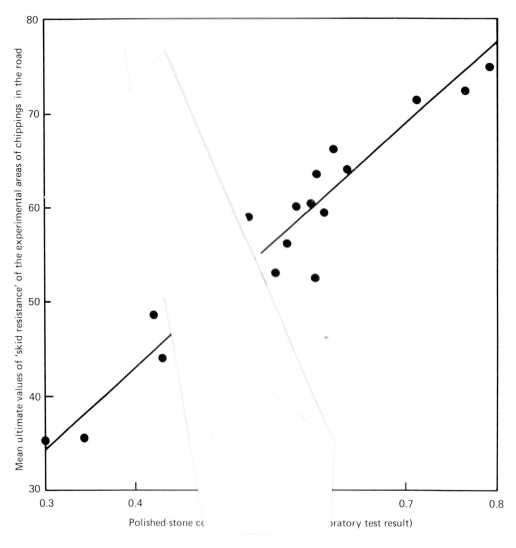

Fig. 25 *Correlation between the results of the laboratory test and the ultimate state of polish of chippings on a straight heavily trafficked road.*

5.1.1.4 Polished-stone value test

In 1960 the accelerated polishing test was published by the British Standards Institution (BS 812:1960), the results of the test were expressed as polished-stone values (PSVs) rather than the earlier polished-stone coefficients. These were the coefficient multiplied by 100 with the intention of avoiding any possible confusion with SFC. Later changes in the test procedure introduced in 1965 led to a general reduction in PSV of approximately 10 per cent.

The latest revision of the PSV test is that of BS 812:Part 114:1990. The test is carried out in two stages - the accelerated polishing of the test specimens followed by the measurement of their state of polish by means of a friction test.

Four curved test specimens are prepared from each sample undergoing test. Each consists of 35 to 50 representative chippings of carefully controlled size supported in a suitable rigid matrix (sand/cement or resin).

Fourteen specimens are clamped around the periphery of the "road wheel" and subjected to two phases of polishing by wheels with rubber tyres. (The original test employed pneumatic tyres but the amended procedure employs solid tyres.) The first phase is of abrasion by a corn emery for three hours, this is followed by three hours of polishing with an emery flour.

The degree of polish of the specimens is then measured by means of the portable skid-resistance tester (using a special narrow slider, shorter test length and supplementary scale) under carefully controlled conditions. Control specimens are used to condition and check the slider before the test; also a pair of control specimens is included in each test run of fourteen specimens to check the entire procedure and to allow for adjustment of the result to compensate for minor variations in the polishing and/or friction testing. Results are expressed as "polished-stone values" (PSVs), the mean of the four test specimens of each aggregate.

A study of the precision of the PSV determination (Tubey and Jordan, 1973) showed that the correction technique based on control specimens improved both repeatability and reproducibility. This technique was incorporated in BS 812:1975.

5.1.2 Aggregates and polishing

The distribution of test results from 292 samples of aggregates was studied (Director of Road Research, 1960) and Fig. 26 shows the wide range of values found. In order to obtain a better understanding of this problem petrological examination studies were made of samples from 76 different sources (Knill, 1960). The petrographic origins of aggregates with high and low resistance to polishing were found to differ for the various rock groups. Generally, it was found that rocks composed of minerals of widely different hardness, and rocks that wear by the pulling out of mineral grains from a relatively soft matrix, had relatively high resistance to polishing. Conversely rocks consisting of minerals having nearly the same hardness wore uniformly and tended to have a low resistance to polishing.

A study was also made of the factors that affect the results of PSV tests (Hosking, 1968). These factors can be separated into those connected with the test itself and those arising from the production of the aggregate and from sampling. Recommendations were made for the amendment of the test procedure to improve its reproducibility. These changes were subsequently adopted by the British Standards Institution. In addition, an improved method of manufacturing specimens for the PSV test was developed that overcame the delay that resulted from the long curing time (7 to 14 days) required for the traditional method (Hosking and Szafran, 1968). Use was made of the rapid hardening properties of synthetic resins, this method was also found to be of value in preparing specimens for the aggregate abrasion test.

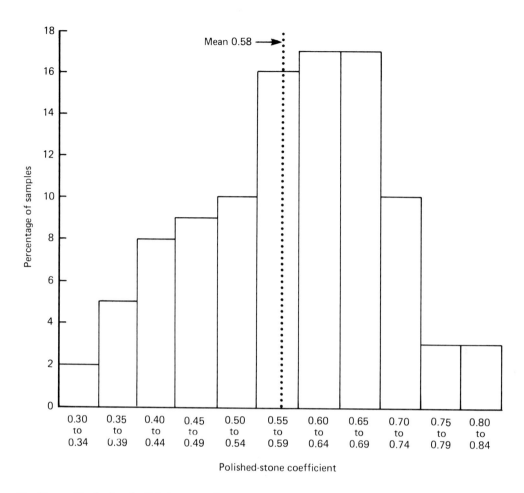

Fig. 26   Distribution of polished-stone coefficients of roadstones from 292 sources.

5.1.3   The effect on skid-resistance of factors other than PSV

A study (Hosking, 1973) of aggregate characteristics thought to affect the skid-resistance of bituminous roads showed that, apart from PSV, skid-resistance can be affected by aggregate type, nominal size, grading and durability. It was also found that mixtures of two or more roadstones give results directly proportional to the means of those given by the constituents when they are used alone (see also 5.1.4 below). The main findings were:

i   In general, quartzites and blastfurnace slags gave a resistance to skidding equivalent to that given by other roadstones which were 3 units higher in PSV. None of the other roadstone types studied (basalts, granites, porphyries and gritstones) showed an important departure from the average. (*Author's Note:* the quartzites studied were dense in character with PSVs of about 55; more loosely cemented quartzites can be expected to behave similarly to gritstones).

ii   Providing adequate surface texture is maintained, reducing the nominal size of an aggregate raises the resistance to skidding of a surfacing made with it. Over the range of size studied (25 mm to 3 mm nominal sizes), halving the size of chippings increased the resulting sideway-force coefficient (SFC) by about 0.08 units. In the case of coated macadams the corresponding increase was less (about 0.03 units).

iii   There was some indication that the use of more nearly single-size chippings in surface dressings gave a better surface texture.

iv   Inadequate durability (resistance to abrasion, cracking and weathering) in an aggregate can affect the resistance to skidding.

5.1.4 Mixtures of aggregates

The possibility that a mixture of two or more aggregates of suitable characteristics might give a greater skid-resistance than either on their own has been explored on several occasions. An early study (Hopkins, 1959) showed that 50:50 mixtures of two aggregates gave results in the accelerated polishing test (now the PSV test) that were mid-way between those for the individual aggregates. Later studies (Hosking, 1973) led to the same conclusion; these showed that mixtures of two or more roadstones in road surfacings were found to yield a resistance to skidding and a depth of surface texture that were approximately equal to the means of those given by the constituents on their own. (*Author's Note:* these studies were of good quality road surfacing aggregates. It has been reported that a mixture of aggregates of very different abrasion resistance can give an enhanced macro-texture (coarser road surface texture)).

5.1.5 Relationship between polished-stone value and skidding

The Transport and Road Research Laboratory and a number of other organisations have carried out a considerable number of small-scale and full-scale road experiments which have provided data of value in studying the relationship between PSV and the skidding resistance achieved in the road (e.g. Wilson, 1966; Brown, 1967; Hosking, 1967b).

Although significant correlation between PSV and skid-resistance was found in each of these road experiments, the relationship was found to vary from site to site. Examples are shown in Figs. 27-30. At first it was believed that this was due to some sites taking longer than others to reach an "ultimate" state of polish. Later it was realised that factors other than PSV were also influencing resistance to skidding and it was an "equilibrium" state rather than an "ultimate" state that was being reached.

The availability of SCRIM in the late 1960s increased the amount of data on skid-resistance so that multiple regression techniques could be used to study the effect of PSV and other factors on resistance to skidding (Szatkowski and Hosking, 1972). A highly significant multiple correlation was found between SFC, PSV and commercial traffic density (correlation coefficient of 0.91 for 139 observations) which took the form:

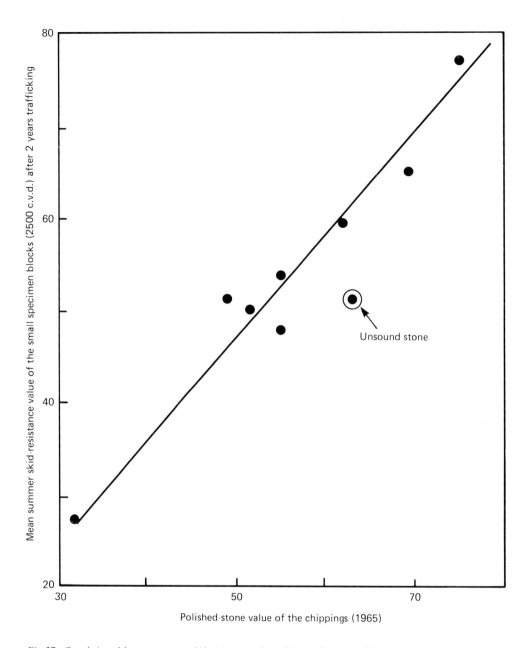

Fig. 27   Correlation of the mean summer skid-resistance values of the small specimen blocks with the PSV of the chippings.

$$SFC = 0.024 - 0.0000633 CVD + 0.010 PSV$$

where

CVD is the number of commercial vehicles per day.
PSV is the polished-stone value.

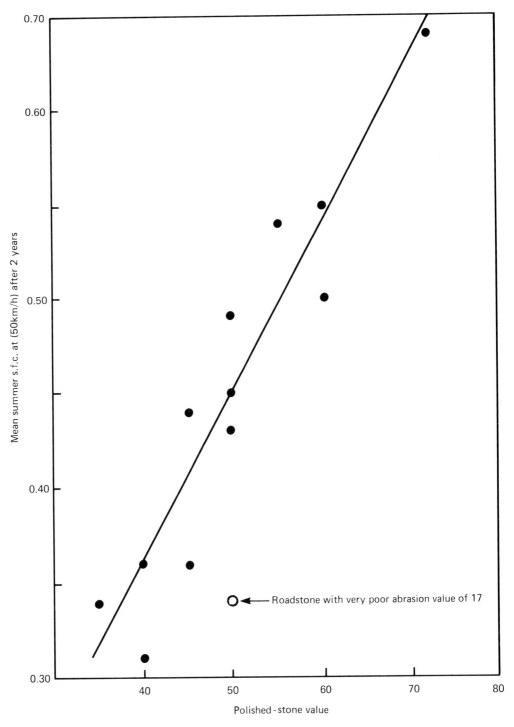

Fig. 28 Correlation between polished-stone value and sideway-force coefficient for 12 mm chippings in tar surface dressings.

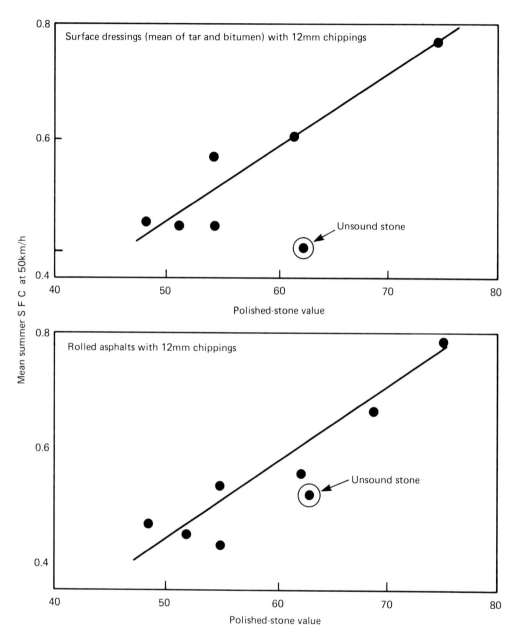

Fig. 29   Correlation between the polished-stone value of the stone with the mean summer SFC of the road surface after 4 years trafficking.

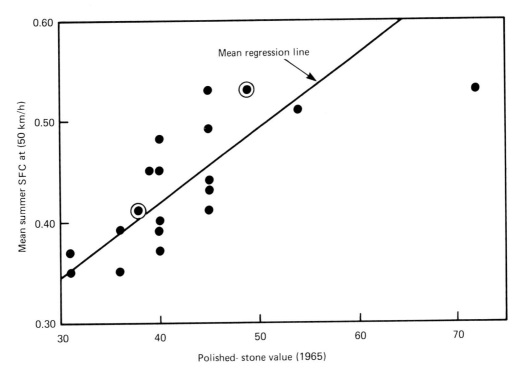

Fig. 30   Correlation between sideway-force coefficient at 50 km/h and polished-stone value (1965) of 19 mm chippings in rolled asphalt.

It was also found that the SFC could be estimated reasonably accurately from total traffic. This analysis also showed that the particular SFC level of a site was reached within about 12 months of trafficking (Fig. 31), after which the mean summer SFC remained virtually unchanged for many years unless there was a change in traffic. Fig. 32 also shows this phenomenon and, in addition, demonstrates that the traffic effect is reversible. In this example a section of heavily trafficked road was by-passed by the opening of a motorway, and the polished surface became more skid-resistant under the ensuing lighter trafficking.

Later work on urban roads in Greater London (Young, 1985) yielded a rather smaller benefit in SFC resulting from an increase in PSV, *viz.* a factor of 0.008 rather than 0.010. The lighter rate of spread of chippings on London's urban roads at the time was given as a possible reason for this difference.

(*Author's Note:* This book is mainly concerned with the part played by aggregates in determining the resistance to skidding of road surfacings. Other factors can significantly affect the skid-resistance of flexible road surfacings. These include the weathering properties of the binder and the mix design itself. However, important as they are, their effect is generally small compared with the difference between a high- and a low-PSV aggregate.)

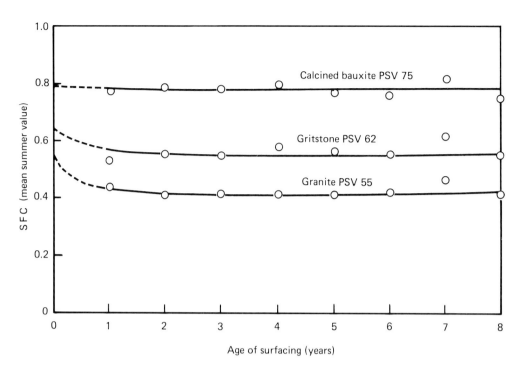

Fig. 31   Levels of skidding resistance recorded on different sections of the same road.

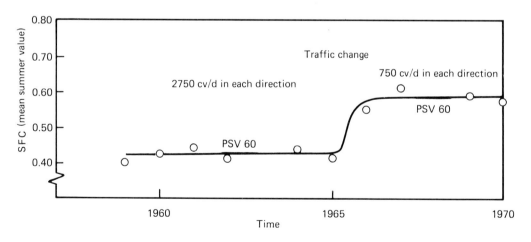

Fig. 32   An increase in the level of skid-resistance recorded on Trunk Road A4, Colnbrook By-Pass, when traffic decreased due to the opening of a motorway.

5.1.6 Variation in skid-resistance at a road site

Skid resistance can vary over quite small distances. An example is the transverse distribution of skid-resistance across a road, where lowest values are obtained in quite narrow wheel-tracks and much higher values are found elsewhere. Fig. 33 shows the distribution of skid-resistance that was found on a quadrant of a roundabout on the trunk road A38 at the Almondsbury interchange; in particular, it shows that lower-than-average values are given where the traffic is turning or braking (see 5.1.7 below).

5.1.7 Turning and braking

An account of the study of the extra polishing of aggregates in bituminous surfacings that takes place at sites where traffic is turning and braking has been published (Hosking and Tubey, 1974). The sites studied included bends with radii varying from 40 metres to 250 metres, approaches to roundabouts and approaches to pedestrian crossings. Examples of the results are given in Figs. 34 and 35. It was concluded that aggregates to be used in surfacings at most sites where traffic was turning or braking should have a PSV at least 5 units higher than that for similar event-free sites, in order to maintain the same level of resistance to skidding. This requirement is additional to the extra resistance to skidding that may be necessitated by any additional potential hazard of the site.

5.2 Aggregates in concrete

5.2.1 Accelerated wear machine

The Transport and Road Research Laboratory and the Cement and Concrete Association made a collaborative study of factors affecting the resistance to skidding of concrete roads (Weller and Maynard, 1970a,b; Weller, 1970). Because of the lack of data from road sites they designed an accelerated wear machine to simulate the trafficking on the road within the laboratory site. It was found that the most influential constituent in the mix was the fine aggregate, and that the characteristics of the coarse aggregate had only a slight effect on resistance to skidding. This work led to the adoption by the Department of Transport of the accelerated wear test for assessing concretes made with limestone aggregates (Department of Transport, 1988).

Subsequent studies were made of data obtained under actual road conditions by means of two core-insert experiments (Franklin and Calder, 1974). The sites chosen were on Motorway M4 between Chiswick and Langley and on the trunk road A12 at Brentwood. Their findings confirmed that it is the fine aggregate that is the most important constituent of a mix in determining skid-resistance, and that it should have a high resistance to both abrasion and polishing. Other factors that were found to affect skid-resistance were the sand content, the strength of the concrete and the hardness of the coarse aggregate (softer aggregates tended to give better results).

Fig. 33   Distribution of skid-resistance values on a segment of the Almondsbury Interchange.

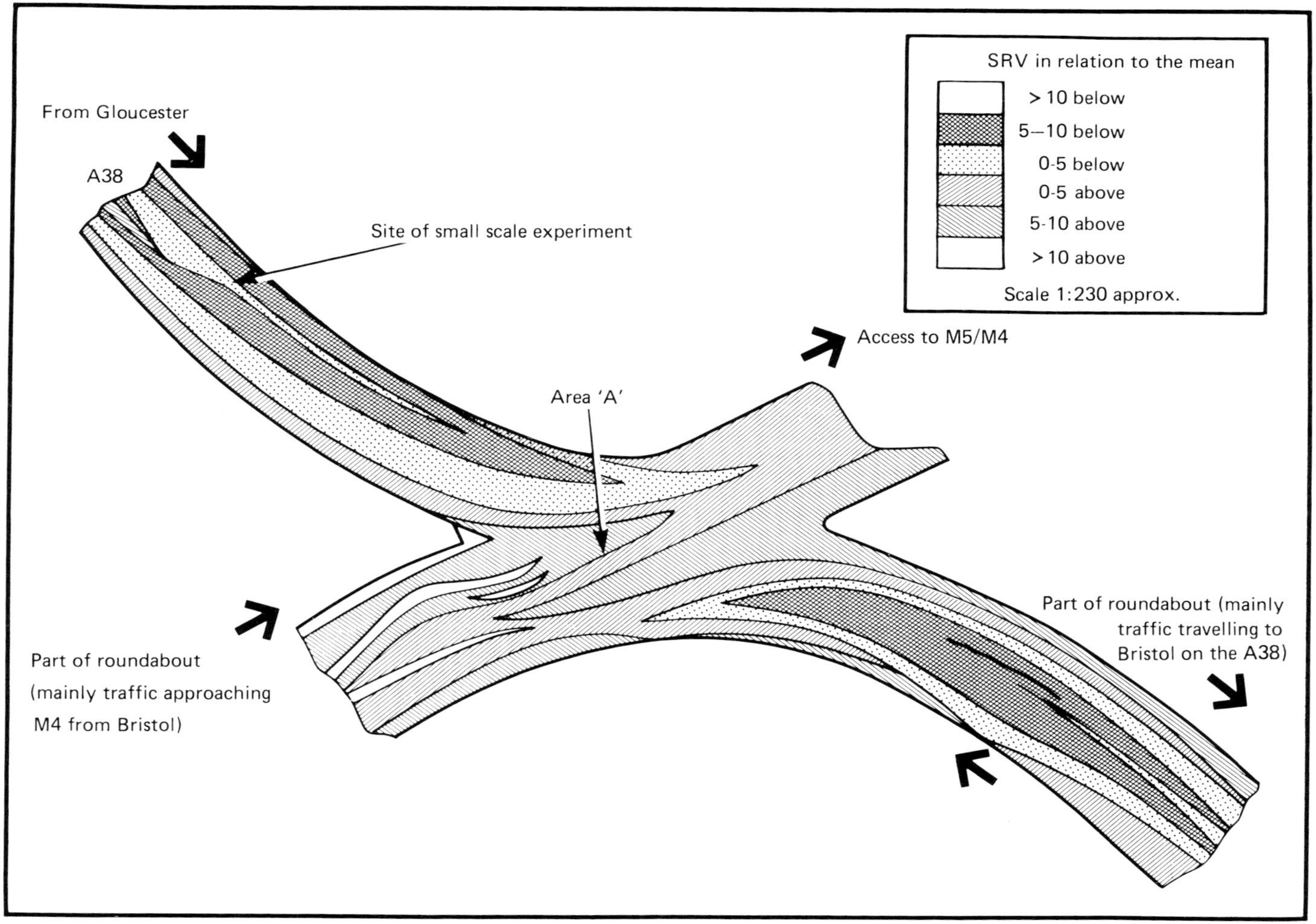

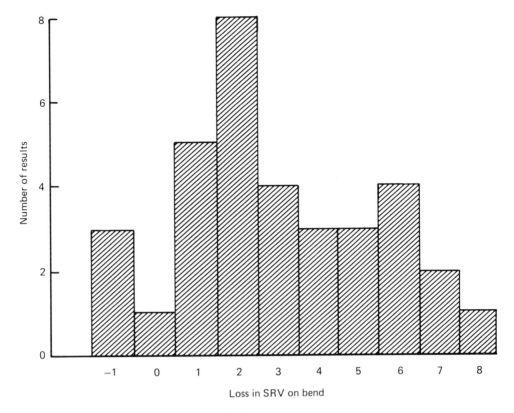

Fig. 34   *The distribution of loss in SRV on bends.*

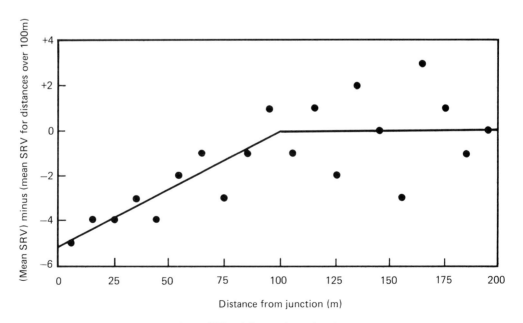

Fig. 35   *Relation between the loss in SRV and distance from a junction.*

5.2.2 The polished-mortar value test

Having established that the major material factors influencing the low-speed skidding characteristics of a concrete road were the resistance to polish and abrasion of the fine aggregate, the TRRL developed a new method (Franklin, 1978) to measure directly a combination of these properties in the polished-mortar value (PMV) test.

This test was an adaptation of the method and apparatus used for measuring the polished-stone value (PSV). Samples of the fine aggregate were mixed with ordinary Portland cement to produce a mortar which was used to cast specimens similar to those used in the PSV determination. These were then subjected to a polishing and friction test procedure similar to, but not the same as that used in the PSV determination. Resultant values on a range of fine aggregates were found to be as much related to the hardness of the aggregate as to the PSV of the corresponding coarse aggregate. Correlations were sought with the results of two concrete core-insert experiments, on the A12 and M4 respectively, where a wide range of fine and coarse aggregates had been used and subjected to traffic for up to seven years. These revealed significant linear correlation between the skid-resistance values of the trafficked surfaces and the results of the new test, designated as polished-mortar values (PMVs). (See Figs. 36 and 37.)

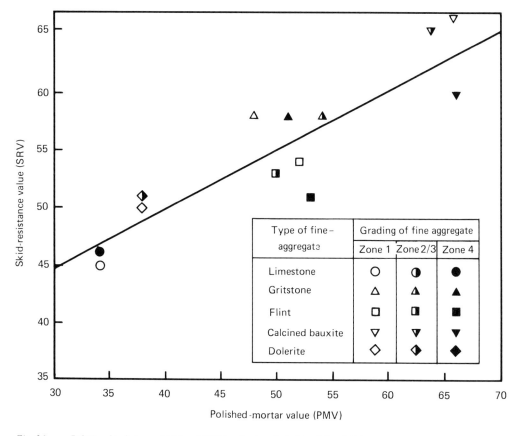

Fig. 36   Relationship between PMV and SRV in wheel-paths in Trunk Road M4.

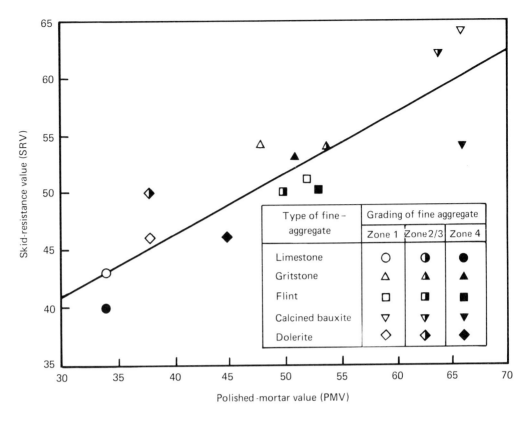

Fig. 37   *Relationship between PMV and SRV of core inserts in Trunk Road A12.*

Further examination of the data, using multiple regression analysis, confirmed that the PMV determination on the fine aggregate adequately described the primary factors of materials influencing the low-speed skidding resistance of concrete road surfaces. Secondary factors which were identified were the fine-aggregate content and, on the very heavily trafficked M4 site, the PSV of the coarse aggregate. However, the size of the contributions made by these secondary factors was relatively small.

The multiple regression equations were:

i   for the A12 data, $SRV = 15 + 0.60 PMV + 0.24 FAC$
ii   for the M4 data, $SRV = 6.1 + 0.61 PMV + 0.25 FAC + 0.11 PSV$

where

> SRV is the skid-resistance value determined by the portable skid-resistance tester
> PMV is the polished-mortar value determined by the new procedure
> FAC is the fine-aggregate content (percentage by mass of aggregate passing a 5 mm BS sieve)
> PSV is the polished-stone value

141

5.2.3 Black deposits

Black deposits have appeared on heavily trafficked sections of motorways after spells of dry weather. They reduced the resistance to skidding and were also thought to contaminate streams after subsequent rain. The problem was studied (Green, 1974) and the deposits were found to consist mainly a mixture of oil, tyre-rubber and dust. The basis of formation of these deposits was found to be the lubricating oil dropped by vehicles. It was considered that prevention rather than remedial action would be the more effective method of dealing with the problem. Contacts were established with the motor industry and modifications were discussed which would reduce the degree of oil spillage, both from poorly maintained sources (sumps, gearboxes, etc) and from automatic chassis lubrication systems which are designed on the total-loss principle. The problem has now largely disappeared as vehicles have been re-designed.

5.3 Aggregates in other road materials

5.3.1 Resin surfacings

5.3.1.1 General

An outline of the development of this method of achieving an exceptionally high resistance to skidding is given in Chapter 3.1.4

Following the early success with resin-bound surfacings on the Colnbrook By-Pass in 1959 and successful trials on bridge decks, a calcined-bauxite/epoxy-resin (CBER) dressing was successfully used on Haven Bridge at Yarmouth in 1961. The dressing was carried out in situ and suffered to some extent because of wetness of the wood-block bridge deck at the time of laying. In 1962 a similar dressing was used on Tower Bridge, London in the form of pre-treated wood blocks to avoid the problem of wetness (James, 1963). It is interesting to note that the re-paving of this bridge with a light-weight polyurethane base covered by the dressed wooden blocks enabled it to support the increasing traffic loads. A very high skidding resistance was maintained on this bridge for a very long time. Many other bridges have been treated similarly with CBER dressings, in order to obtain a highly skid-resistant surface with little increase in weight.

In the mid-1960s, Hatherley of the Greater London Council's road department observed that about three-quarters of traffic accidents took place on urban roads, London's roads constituted a high proportion of the country's urban roads, and that a considerable proportion of London's road accidents occurred at a relatively small number of "black-spots". This led him to believe that the special treatment of relatively few road sites would lead to a significant reduction in the number of the country's skidding accidents. After consulting the Road Research Laboratory as to the best means of achieving a very high skid-resistance, Hatherley considered that very worthwhile reduction in accidents could be achieved by resurfacing the black spots with CBER surface dressings.

Collaborative work by Greater London Council, TRRL, Shell-Mex & BP Ltd., Prismo Universal Ltd and the London Borough of Lambeth led to a number of trials being laid on roads in London in 1967 with a CBER dressing system known as "Shellgrip" (Hatherley and Lamb, 1970). The sites were selected because of their high risk of accidents. After the treatment the reduction in accidents was spectacular and led to the widespread use of this and similar surfacings in London and, later, elsewhere. The cost of the treatment was high but it had the advantage of being able to be laid during a few hours overnight and so cause few delays to traffic. An additional advantage was that because it was a thin surfacing, it could be applied without the need to excavate existing material.

The success of the CBER dressings is considered to be a result of several contributing factors:

i   The high polishing resistance of RASC-grade calcined bauxite (PSV = 75).

ii   The small size of the "chippings" (2.8 mm to 1.2 mm).

iii   The "edge-on" effect of the chippings which are firmly fixed in the rigid binder. This contrasts with thermo-plastic binders such as bitumen, where chippings tend to be orientated into the most stable position - with a face rather than an edge at the surface.

iv   The rigid nature of the binder which resists embedment of the chippings by traffic.

Junctions in London that had been surfaced with "Shellgrip" maintained an exceptionally high resistance to skidding over many years. Fig. 38 shows the high sideway-force coefficients measured during the first four years of their life.

5.3.1.2 Aggregate requirements

A study (Hosking and Tubey, 1972) of the performance of a wide range of aggregates in resin surfacings was made by the TRRL in collaboration with the Greater London Council, Shell Mex and BP Ltd, Shell Research Ltd. and Prismo Universal Ltd. The results showed that the standard BS 812 tests for aggregates were not suitable for assessing aggregates for use as 3 mm grit. However good performance was found to be associated with aggregates with a minimum PSV of about 70 combined with an aggregate abrasion value of 5 or less.

(*Author's Note:* It must be stressed that considerable damage to the abrasion lap can occur when testing calcined bauxite and other very hard materials, and so such a test is not recommended for such materials. Tests with a Taber "Abraser" (used in some American standard test procedures) were encouraging, but more study is needed before recommendations can be made.)

Fig. 39 shows the correlation that was obtained between PSV and mean "road rating" (a measure of resistance to skidding based on skid-resistance value). Aggregate impact tests on 3.2 mm to 2.4 mm grit were found to yield meaningful results when the fines were separated with a 600 μm test sieve, rather than the standard test sieve. A value of 25 or less was found to indicate a satisfactory strength performance.

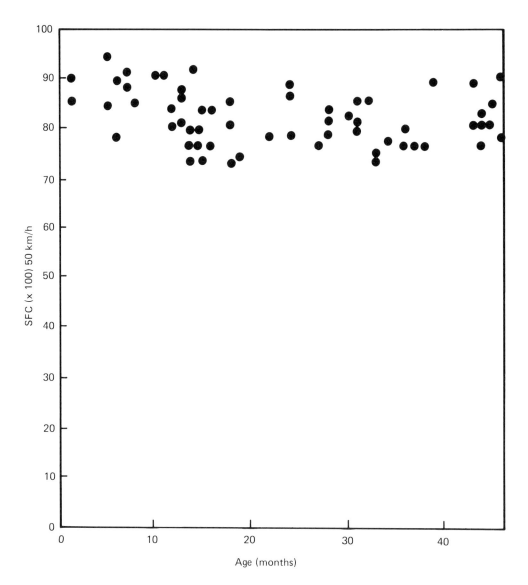

Fig. 38 Performance of "Shellgrip" at road junctions.

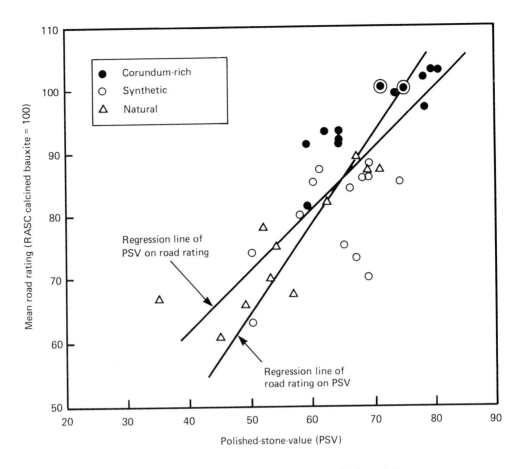

Fig. 39   Relation between the PSV and road rating of aggregate in resin-binder surfacings.

5.3.2   Other special polish resistant surfacings

5.3.2.1   "Delugrip"

Delugrip is a proprietary surfacing material developed by Lees of the University of Birmingham and Dunlop Ltd. This material is the product of a "rational design method" for bituminous mixtures (a method has also been developed for concrete). The object is to obtain a durable surfacing material which is both quiet and skid-resistant. A feature of the design is the use of aggregates of very different abrasion resistance so that the macro-texture is maintained by their differential wear. (See also 5.1.4 above).

Delugrip has been the subject of a number of trials in London and on other roads. At many sites it has been reported to have yielded values of sideway-force coefficient that were about 0.10 units higher than would have been expected with a comparable hot-rolled asphalt (Young, 1985).

A wide range of bituminous surfacing materials have been studied in a full-scale road experiment on Trunk Road A1 near Buckden. These included sections of open-textured macadam, pervious macadam, dense bitumen macadams, rolled asphalts, surface dressing and five sections of Delugrip surfacings. After five years of trafficking (approximately 2,000 commercial vehicles per day) it was concluded (Brown, 1986) that all the mixed materials showed adequate resistance to deformation, produced a good riding surface and provided a higher level of resistance to skidding at low speeds than the recommended target value. However those materials with a low texture depth, including Delugrip surfacings, showed a considerable fall-off in resistance to skidding at high speeds.

5.3.2.2 Friction courses and pervious macadams

The Air Ministry developed friction courses to reduce aquaplaning on airfield runways. The principle was to provide a permeable surfacing through which rain water would drain away. The principle has since been applied to roads with the object of providing good high-speed skidding characteristics combined with low tyre noise and reduction in splash and spray. These pervious macadams are laid over an impervious layer, water drains through the pervious material and thence to the side of the road. Trials in Great Britain have been carried out at a number of sites in Warwickshire, Staffordshire and London (e.g. Brown, 1973; Brown, 1977; Vallis, 1978: Young, 1985; Daines, 1985).

5.3.2.3 Slurry seals

Slurry seals consist of a mixture of fine aggregate, filler, bitumen emulsion and water. Although not normally considered to be suitable for surfacing heavily trafficked roads, a proprietary slurry seal has given promising results in trials in London (Young, 1985). (See also Chapter 5.5.3.7.)

5.3.2.4 Exposed aggregate concrete

It is the nature of the fine aggregate that determines the skid-resistance of a conventional concrete surfacing, but if the coarse aggregate is exposed at the surface then its polishing characteristics will become important, as in most bituminous materials. Exposed aggregate concrete has been used for decorative purposes for many years. More recently this type of surfacing has been studied, particularly in Belgium, with the object of providing durable skid resistant concrete surfacings (Leyder, 1965). Special mechanical equipment has been designed that will lay and brush away the superfluous mortar from the surface before it sets. This is done within three hours of laying by wetting the surface and brushing with flexible nylon bristles. Road trials at Herent in Belgium showed that a high and durable skid-resistance was obtained when using high-PSV aggregates and that SFC values were proportional to the PSV of the coarse aggregate.

5.3.2.5 Chipped concrete

The chipping of concrete is a new type of process that has been developed in Belgium (Heystraeten, 1975). It should not be confused with the surface dressing of concrete (See Chapter 3.1.3.4). Polish-resistant single-sized chippings are spread over the surface of freshly laid and levelled concrete. They are then sprayed with a curing agent and embedded by means of a vibrating hammer, so that they are slightly proud of the concrete surface.

This process provides good macro-texture and at the same time imparts a skid-resistance that is determined by the choice of aggregate in the same way as in bituminous rolled asphalt surfacings.

Road trials on the Hasselt By-pass in Belgium have shown that the skid-resistance varies with the PSV of the chippings and that the resistance to polishing of stone in the body of the concrete does not affect resistance to skidding.

A similar process has been used in Canada (Ryell, Hajek and Musgrove, 1976) where the chippings were spread before compaction of the concrete and the surface laitance removed by brushing on the day after laying. In this respect it resembles exposed aggregate concrete discussed above.

5.4 Mechanism of polishing of aggregates

The degree to which aggregates polish in the road surface has been found to depend on the type of aggregate, the intensity of trafficking and weather conditions. Studies of this mechanism have been made by a number of workers including Neville at the TRRL (Neville, 1972; Neville, 1974). She developed a technique which allowed the continuous (non-destructive) observation of roadstones in a road surface at magnifications of up to x5000 and applied the technique to the study of the behaviour of a range of roadstones at two sites.

This work showed that two ranges of micro-texture (100 µm to 250 µm and <10 µm) had important effects on the low-speed resistance to skidding of roadstone exposed to trafficking, and that each was affected by season of the year, traffic density and the mineralogy of the roadstone. Four mechanisms were found to be involved: polishing, abrasion, differential wear and weathering. Apart from general smoothing, the observed state of polish was sometimes the highly glossy condition that is achieved when Bielby flow takes place (i.e. the surface is burnished by the heat and pressure of friction). This condition was observed after dry road conditions in summer, but was seen to be lost during the following wet winter conditions as a result of abrasion. Some roadstones showed additional roughening during the winter months as a result of differential wear and weathering.

It was concluded that the main factors affecting the degree of skid-resistance given by aggregates in the road surface are the traffic density, the range of hardness of the minerals present in the aggregates and their susceptibility to polishing. It was also observed that the quartzite studied showed virtually no seasonal variation in micro-texture (the fine surface texture that affects the PSV of an aggregate), thus offering an explanation of why this group of aggregates give a rather better resistance to skidding than most aggregates of comparable polished-stone value.

5.5 Road experiments

5.5.1 General

The TRRL and a number of other organisations have carried out a considerable number of small-scale and full-scale road experiments which have provided data of value in studying the relationship between laboratory tests and road performance. An example is the relation between PSV and the skidding resistance achieved in the road.

Small-scale road experiments and core insert experiments provide a convenient means of relating PSV to the polishing effect of traffic. Panels of chippings are inserted in the surface of the road at selected sites and measurements of skid-resistance are made at intervals with a portable skid-resistance tester. These experiments have the following advantages:

i   They make use of the same instrument for both laboratory and road measurements.
ii   They enable similar samples to be used for the PSV test and in the road.
iii   They eliminate differences due to type and quality of the surfacing.
iv   They permit the removal of the experimental samples for examination in the laboratory and their subsequent replacement.
v   They yield a comparison of the performance of a wide range of aggregates under virtually identical conditions and, in particular, it is possible to carry out trials at difficult sites such as roundabouts.
vi   Values can be safely obtained for aggregates that might prove to provide unsatisfactory road surfaces.

Correlation between PSV and resistance to skidding has been found to be better than with full-scale experiments because of the elimination of many of the variables.

An alternative to the use of small-scale experimental panels has been a core transplant technique. This technique allows experimental mixes to be made under carefully controlled conditions in the laboratory, cores to be drilled from them and inserted into the road, flush with the surface. A rapid hardening resin is used to fix the cores in position. In this way a wide range of materials of controlled composition can be compared under identical road conditions.

Full-scale road experiments consist of lengths of actual road surfacing materials and are usually from 50 metres to 200 metres in length. This length is sufficient to give a reasonably large sample and to allow skid-resistance measurements to be made with testing vehicles such as SCRIM and the BFC trailer (see Chapter 6). They have the following advantages:

i   They allow trials to be carried out under road conditions.
ii   They allow tests to be carried out at traffic speeds rather than at the very low speed of test with the portable tester.
iii   They allow comparisons to be made of aggregates in different types of surfacing.
iv   They allow direct correlation between PSV and SFC.

5.5.2 Results from small-scale road experiments

Small-scale road experiment panels were first used by the Road Research Laboratory in their studies of the polishing characteristics of aggregates in the road in the 1950s (Maclean and Shergold, 1958). Sand/cement mortar panels were surfaced with the aggregate under examination in a similar way to PSV test specimens. These were mounted in sets of four on steel plates that were then bolted in frames so that they were flush with the road surface. This enabled their ready removal for examination and assessment in the laboratory and subsequent replacement. In one of Maclean and Shergold's experiments, sets of 8 different aggregates covering a wide range of aggregate types were set in roads with traffic conditions ranging from a quiet country lane to a heavily trafficked roundabout. The aggregates ranged from a polish-susceptible Carboniferous limestone to a polish-resistant greywacke gritstone. The skid-resistance of the panels was measured at frequent intervals over a period of several years: results after 4 days, 4 weeks and 3 months are shown in Fig. 40. This figure clearly shows the very rapid polishing that took place on the most heavily trafficked site after only 4 days.

5.5.3 Results from full-scale road experiments

5.5.3.1 Haddenham experiment

This experiment, more fully described in Chapter 3.1. was carried out to compare the performance of eight aggregates as chippings on a lightly trafficked road. After several years, none of the eight stones had polished. It became evident that under such light traffic the weaker of the aggregates used, which would have hitherto been rejected by many engineers, would give many years of useful life.

5.5.3.2 Streatham High Road experiment

This experiment together with that on the Derby Ring Road, described below, was carried out to correlate the results of mechanical tests on aggregates with their performance when used as pre-coated chippings in a hot-rolled asphalt surfacing under heavy traffic. The Streatham site was on the A22 trunk road between Pendennis Road and Glenelson Road carrying approximately 40,000 tonnes of traffic at the time of laying (1953). 20 sections each of over 30 metres in length were laid using chippings of two sizes, 1/2inch (12.5 mm) and 1 inch (25 mm). The aggregates used were granite, hornfels, quartzite and blastfurnace slag. A 50:50 mixture of granite and slag was also used.

By 1955 at this site, now carrying 60,000 tonnes per day, all the aggregates showed considerable wear (Director of Road Research, 1955), manifested both as polishing and attrition. The resistance to skidding with 1/2 inch (12.5 mm) aggregates was also found to be consistently higher than with 1 inch (25 mm) aggregate.

5.5.3.3 Derby Ring Road experiment

This experiment was carried out for the same reasons as the Streatham High Road experiment (see above). The road was not as heavily trafficked as the Streatham experiment (17,000 tonnes per day at the time of laying (1953).) The site was on the A5111 trunk road - Harvey Road on the Derby ring road.

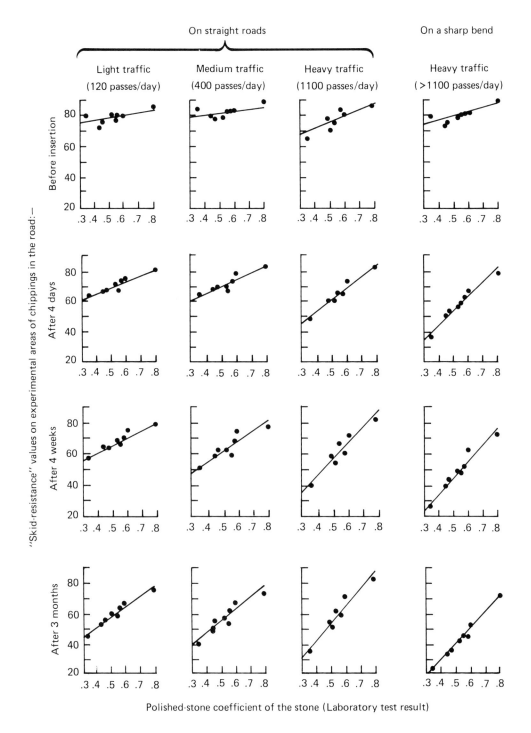

Fig. 40 Correlation between the results of the laboratory test and the rate of polish of experimental areas of chippings.

By 1955 this and the Streatham High Road experiment showed that where aggregates are used as coated chippings on more heavily trafficked roads, resistance to skidding is much lower than on more lightly trafficked roads. Fig. 30 shows the results obtained after 12 years trafficking of this experiment where 17 different aggregates were used as 3/4 inch (19 mm) coated chippings in rolled asphalt (Brown, 1967). Additional sections were laid using mixtures of two types of chippings: their performance was equivalent to the average of their test results. Although the highest PSV aggregate gave the best results, its performance was not as good as expected. This was attributed to the relatively poor resistance to abrasion of this stone coupled with the long period of trafficking.

5.5.3.4 West Wycombe experiment

The West Wycombe experiment was laid in 1955 to assess the performance of different aggregates used for surface dressing a heavily trafficked road, and to correlate the results with the results of laboratory tests. The site was on the London to Oxford trunk road (A40) at West Wycombe in Buckinghamshire. The road was 21 ft (7 m) wide and carried 25,000 tonnes of traffic a day at the time of laying. The thirteen aggregates used were screened to give 70 to 90 per cent of true 1/2 inch (12.5 mm) size chippings, and were laid with three different rates of spread of binder.

Fig. 28 shows the relationship between SFC and PSV that was obtained after two years of trafficking (Wilson, 1966). Good correlation was obtained for the thirteen different aggregates ranging in PSV from 35 to 72 and yielding SFCs ranging from 0.32 to 0.64. One of the aggregates demonstrated the need for adequate abrasion resistance as well as PSV, with an aggregate abrasion value (AAV) of 17 it wore away rapidly under the trafficking and consequently gave a relatively poor SFC. A "life" of nine years was achieved with many of the aggregates, but some had to be replaced before that because of inadequate resistance to skidding. As a result care was taken to exclude the less polish-resistant aggregates from later full-scale road experiments.

5.5.3.5 Blackbushe experiment

Experimental sections of bituminous surfacings were laid in 1962 on Trunk Road A.30 at Blackbushe, Hampshire, using eight different roadstones with PSVs ranging from 49 to 75. The types of experimental surfacings included surface dressing, rolled asphalt, bitumen macadam and dense tar surfacing.

The results of tests carried out over the first four years showed (Director of Road Research, 1966) that the PSV of the stone used in the road surface was the main factor determining the resistance to skidding. There was also some evidence that other roadstone properties (such as adhesion and tendency to wear) could affect the resistance to skidding. The material in which the aggregate was used also affected the resistance to skidding. The main differences were that the macadams made with pitch-bitumen gave a higher resistance to skidding than those with petroleum bitumen, the open-textured bitumen macadams gave a rather higher resistance than the surface dressings and the 1/2 inch (12.5 mm) chippings in rolled asphalts gave a higher resistance than the 3/4 inch (19 mm) chippings. The experiment also showed that calcined bauxite (RASC Grade) could provide a satisfactory road surface, with an exceptionally high resistance to skidding, when used as chippings in rolled asphalts and surface dressings.

TABLE 19

Accidents on elevated length of M4
(Main carriageway only excluding accidents on snow/ice and where it was not known whether skidding occurred)

Class	Dry roads			Wet roads		
	With skid	Without skid	% with skid	With skid	Without skid	% with skid
Before resurfacing:						
Injury	16	59	27	27	42	64
Damage only	30	115	26	61	91	65
After resurfacing:						
Injury	8	43	19	8	17	47
Damage only	13	70	19	14	32	44

Fig. 29 shows that good correlation was obtained between PSV and resistance to skidding for the surface dressings and chippings in rolled asphalt after four years of trafficking (Hosking, 1967b) (compare with the results from the small-scale road experiment carried out at the same site: Fig. 27), and that poorer correlation was obtained with the macadams. The scatter of the points has been exaggerated by the scale and, unlike the earlier experiments, only aggregates of relatively high PSV were used at this site. One aggregate gave disappointing results, its SFC being noticeably lower than was expected from its PSV. Inspection of the surface showed that it had worn badly, and subsequent research showed that this was due to its poor weathering characteristics, again demonstrating that high-PSV alone is not enough.

5.5.3.6 Elevated section of M4 experiment

An example of the way in which highly polish-resistant aggregates have been used to reduce accidents (and traffic problems) was the M4 experiment (Miller and Johnson, 1973). This site, the elevated part of Motorway M4, was found to have a high proportion of skidding accidents when wet. Examination of records showed that the SFC of the surface was from 0.35 to 0.45 at 50 km/h in 1967 and 1968. The road was re-surfaced with the highest PSV materials available. It had been intended to use calcined bauxite (RASC Grade) for the entire site, but as insufficient material of the necessary chipping size was available it was decided to mix it with the next highest PSV material available (a gritstone from Gilfach quarry near Neath, with a PSV of 71) for some of the sections: the remainder were re-surfaced with the gritstone alone.

During the first three years after resurfacing the SFC was found to have been increased to between 0.50 and 0.60 and accidents were substantially reduced. Details are given in Table 19.

These figures show that the improvement in SFC of about 0.15 units yielded a reduction in wet skidding rate of 17 per cent for injury accidents and 21 per cent for damage only accidents.

5.5.3.7 Sandy experiment

To examine the performance of calcined-bauxite/epoxy-resin (CBER) surfacings at higher speeds, the braking-force coefficient (BFC) of a 200 m length on A1 Trunk Road at Sandy, Bedfordshire, was regularly measured. The results showed that there was little loss of resistance to skidding at speeds up to 130 km/h and that high levels of SFC were maintained for several years (Fig. 41) (James, 1971; James and Lamb, 1974). Two sections of slurry seal containing calcined bauxite aggregate were also found to be satisfactory after 15 months traffic on the A1 trunk road at the approach to a roundabout.

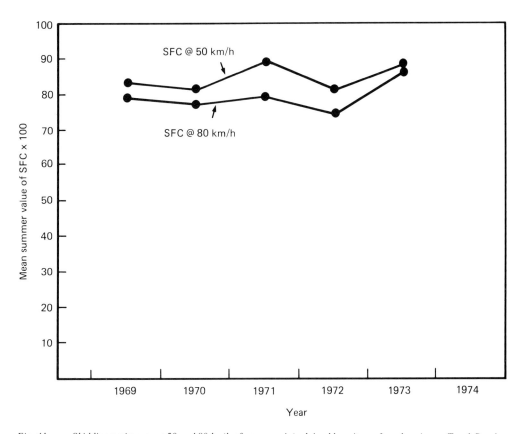

Fig. 41   Skidding resistance at 50 and 80 km/h of epoxy resin/calcined bauxite surface dressing on Trunk Road A1 at Sandy.

6 Road surface characteristics

6.1 Resistance to skidding

6.1.1 General

Skidding occurs when the coefficient of friction between the tyre and the road is inadequate to maintain full adhesion. Although many factors contribute to the occurrence of skids it is the part played by the characteristics of the road surfacing material that forms the main subject of this chapter. The skidding problem cannot be satisfactorily considered in isolation as it is the optimisation of the characteristics of both the tyre and the road that is necessary in order to reduce the risk of a skid occurring.

When roads are dry and free from loose and other contaminant materials the friction between tyre and road is, almost without exception, high. This means that the available friction permits full use to be made of the braking and cornering performance of the vehicle. At the other extreme, ice or snow covered roads have low friction and the accelerations and decelerations used in driving gently can generate forces in excess of those available between tyre and road and skidding occurs.

The wet road condition presents a much more complicated situation; the effect of water on the road greatly influences the design of both road surface and tyre. Wet-road friction varies with many factors including the characteristics of the surfacing, the characteristics of the tyre, the degree of wetness of the road, and the vehicle operating conditions. Each of these factors has been studied in depth and those relating to the road surfacing, and in particular the aggregate in it, form a major part of this book and are dealt with in other chapters. However the other factors are equally important and need to be understood if optimum conditions for preventing skidding are to be achieved.

Studies have concentrated on the wet-road skid-resistance of roads. Unless stated otherwise all references to skid resistance in this chapter relate to the wet-road state.

Three important characteristics of the road surface that relate to wet-road skid-resistance are:

i   Adequate micro-texture of the surface.
ii Adequate drainage of the water from the surface.
iii Good macro-texture of the surface, which is needed to maintain skidding resistance at higher vehicle speeds and to enable low-resilience tyres to improve braking performance.

Fig. 42 illustrates the two types of texture that are commonly used to describe road surfacings. Micro-texture is the fine texture that occurs on chippings and other exposed parts of surfacings, whereas the coarser texture created by the interstices between the chippings, or by the grooving in the case of concrete, is termed macro-texture. The deterioration in skid-resistance that usually occurs as vehicle speed increases is illustrated by the results shown in Fig. 43 for the M40 High Wycombe By-pass experiment (Jacobs, 1983).

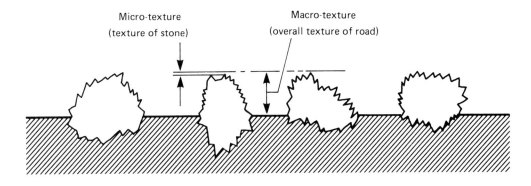

Fig. 42   Micro- and macro-texture of a road surface containing roadstone chippings.

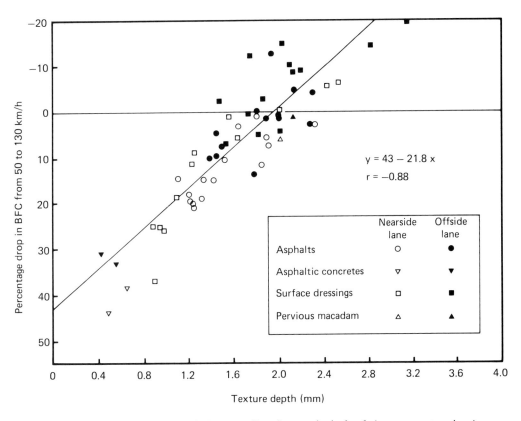

Fig. 43   Motorway M40 High Wycombe By-pass: effect of texture depth of surfacing on percentage drop in resistance to skidding from 50 to 130 km/h.

Methods of monitoring the skid-resistance of road surfacings, such as by the use of SCRIM, need to employ a standard rubber composition and profile in the tyre tread. Natural rubber formed the basis of the compositions used for vehicle tyres at the time that test methods were being developed and consequently the standard was of this type of composition. Fig. 44 shows the wide range of frictional values that were given by different selected rubbers (Sabey and Lupton, 1964), and underlines the need for a standard composition.

Although there have been subsequent changes in the compositions used in tyres for road vehicles, the original test composition has been retained. This was done in order to achieve a continuity

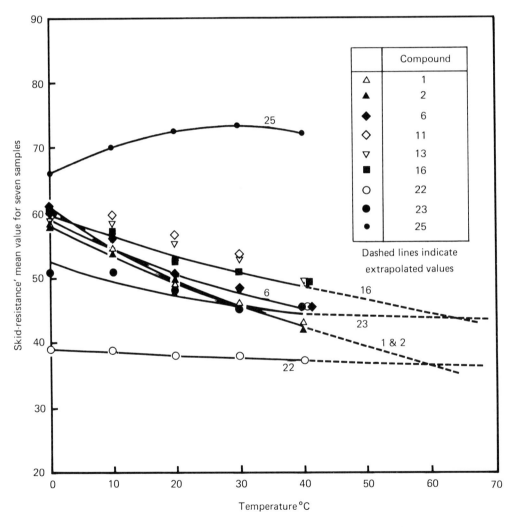

Fig. 44   *Variations in "skid-resistance" with temperature for nine selected rubbers.*

of results and to obtain values that related to the worst case - vehicles still using natural rubber. Additionally, in routine test equipment such as SCRIM, the use of low resilience rubber in the test tyre would reduce the tyre life to an unacceptable level and would probably introduce errors because of excessive heating of the tyre during test. It was also decided to employ smooth tyres for test purposes because they would both relate to the worst conditions on the road, and also avoid test errors resulting from the progressive wear of a tread pattern.

Another limiting factor is the test speed. It is both difficult and dangerous to perform high-speed skid testing without closing the road to other traffic. This effectively rules out the routine monitoring of high-speed skid-resistance.

These factors mean that the measurements obtained with routine skid-testing equipment suffer from the following limitations:

i   They are insufficient to relate accurately to the skid-resistance under higher-speed conditions
ii They cannot be used to assess the effect of the different tyre compositions and tread patterns found on the road.

In practice it has been found that the high-speed skid-resistance characteristics of a road can be satisfactorily assessed by monitoring the macro-texture in addition to the 50 km/h skid-resistance of the road surface.

The measurement of macro-texture is also of value in assessing the benefits achievable by the use of the low resilience (high-hysteresis) rubbers that are now used for the tyres for most light vehicles. Fig. 45 shows the relationship between the coefficient of friction and speed for high and low resilience rubber tyres and for two road surfacings (Lupton, 1968). One surface was smooth (low macro-texture depth) and the other rough (high macro-texture depth). It is clearly shown that the frictional advantages of the low resilience rubber tyres are only achieved with the higher texture depth. This means that, to achieve optimum skid-resistance, adequate macro-texture is needed to maintain skid-resistance at higher speeds, to allow vehicles to make the best use of the better skid-resistance properties of low resilience tyres.

6.1.2   Road surface projection shape and friction

High skidding resistance on wet roads is associated with the presence of sharp edges in the road surface: on these sharp edges high pressures are set up which assist in breaking through the lubricating water film between the tyre and the road. Studies have been made to determine the dependence of skidding resistance on these localised pressures. The pressure distributions beneath rigid spheres and cones pressed into rubber were calculated and experimental evidence confirmed that the rubber was behaving as an elastic material (Sabey, 1958).

Fig. 46 shows the relationship between the coefficient of friction and contact pressures that were achieved with a range of spheres and cones and Fig. 47 shows the relation between the pressure and the angle of the cone. It was concluded from this work that:

i   For satisfactory skidding resistance the individual projections in the road must be such that average pressures of the order of 1,000 lb/sq.in. (7,000 kN/sq.m.) are set up on them.

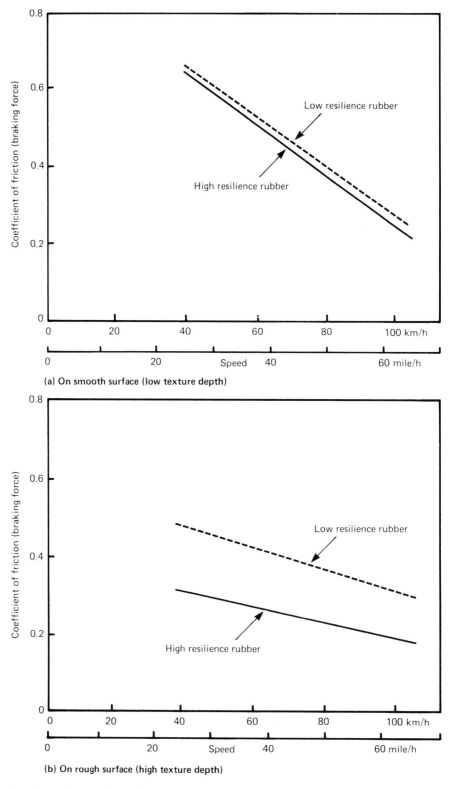

Fig. 45   Tread resilience effect on smooth and rough surfaces (wet).

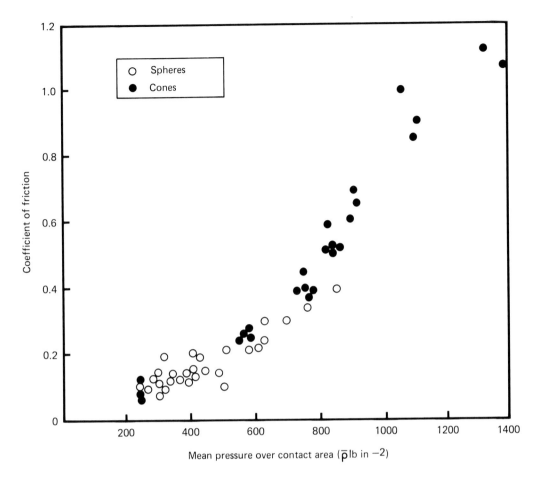

Fig. 46   *Coefficient of friction measured for spherical and conical sliders on wet rubber and the calculated pressure over the contact area.*

ii   The individual projections need to have angles at their tips of 90 degrees or less.
iii The necessary pressures are unlikely to be obtained with rounded or polished projections, whatever their size or the load applied to them.

6.2   Measurement of resistance to skidding

6.2.1   General background to studies in Great Britain

The problem of skidding was recognized in the days of horse-drawn traffic and, for difficult sites such as steep hills, stone setts were selected from quarries where the rock was known to a give better grip. However the problem as we now know it began with the advent of pneumatic tyres

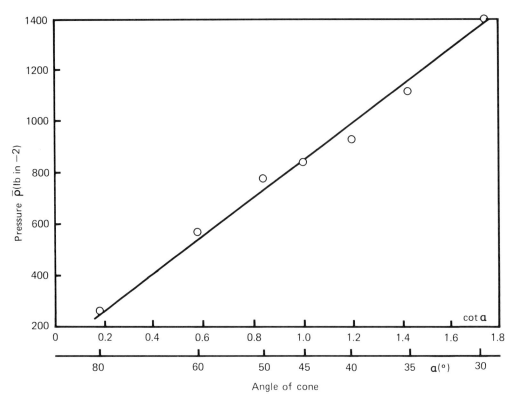

Fig. 47   Pressure and angle of cone.

at the end of the 19th century and was magnified by the big increase in motor traffic after the First World War. Accidents involving skidding became frequent, particularly under wet-road conditions. By 1927 the problem has become so acute that Batson started work at the National Physical Laboratory at Teddington with the aim of obtaining a better understanding of the phenomena involved in the process of a vehicle skidding.

Early experiments into the measurement of the coefficient of friction between rubber and road surfaces soon indicated that the speed of sliding had a marked effect on the results obtained (Bradley and Allen, 1930-1). This led to the development of a specially designed motor cycle and sidecar with the sidecar wheel mounted at an angle to the direction of travel. The ratio of the sideway force exerted by this angled tyre to the vertical force between the tyre and the road provided a means of measuring the slipperiness of the road surface. These forces were transmitted by a link-mechanism to give a single track on a moving pen chart. The ordinate of this record was proportional to the coefficient of friction and so provided the first method of continuously measuring the skidding resistance of a road. The principle of this method of measurement has been used through a succession of test vehicles and is currently used in the most modern method of routine measurement, SCRIM.

Early studies showed that adequate skid-resistance was given by all road surfaces under clean dry conditions. Further work was therefore concentrated on the wet road condition.

The first road experiment designed to study the skidding resistance of a range of materials was laid in 1930 on the Kingston By-Pass. Eight of eleven sections were surfaced with different chippings and the remaining three with porous materials which, it had been claimed, would give adequate wheel-grip. In the event the sideway-force coefficient (SFC) measurements with the motor-cycle apparatus showed that the chippings gave satisfactory results at speeds up to 50 km/h (the maximum speed of test) whereas the lower-textured porous materials showed a marked falling off in skidding resistance at speeds greater than 25 km/h.

Research into skidding was transferred to the Road Research Laboratory (RRL) on its formation in 1933. The virtues of the motor-cycle apparatus were quickly recognized and a more robust version of the apparatus was designed and used for research alongside other methods (Bird and Scott, 1936; Bird and Miller, 1937). The suitability of the SFC method for the routine continuous monitoring of a road was also recognized and a small fleet of motor-cycle/sidecar machines was formed to monitor the country's road network. Unfortunately the advent of the Second World War in 1939 postponed this ambitious project.

Other main types of skid testing were studied at the RRL. The first was the braking-force coefficient (BFC) method that obtained a measure of the coefficient of friction by assessing the ratio of the braking force exerted by a locked wheel mounted on a trailer to the vertical force between the tyre and the road. The method was not suitable for the continuous testing of roads because of over-heating and excessive wear of the test tyre. However it did have the advantage of allowing high speed testing, which was not possible with the SFC apparatus. Some versions of this apparatus allow partial controlled slip of the test wheel so as to widen its applications. The early application of the BFC method was for airfield pavements rather than roads.

Another method makes use of the Izod pendulum principle (Giles *et al*, 1964). The coefficient of friction between a slider and the test surface is measured by assessing the loss in energy in a pendulum apparatus carrying the slider as it slides over the test surface. It is essentially a low-speed apparatus, but it has proved very valuable for laboratory studies; also it has the advantages of low cost and portability. It is usually termed the (portable) "skid-resistance tester" (SRT) and the measurements termed "skid-resistance values" (SRVs). (In some overseas publications these measurements are termed "British portable numbers").

The deceleration of a car with locked wheels has also been used to measure resistance to skidding. A decelerometer fitted in the car is used to measure the deceleration from a speed of 50 km/h when the brakes are applied for a period of one second when all four tyres are locked to make them skid. It has proved to be very useful for comparing the performance of tyre tread patterns. But the practical problems of ensuring the simultaneous locking of all four wheels and the effect of the suspension and tyres on the vehicle make it unsuitable for use on bends and limited its wider usage.

6.2.2 Small braking-force trailer

One of the various braking-force test methods that have been studied by the (Transport and) Road Research Laboratory (T)RRL has been used for studying skid-resistance at high speeds. This small braking-force trailer apparatus was first designed to test airfield runways at high speeds but, because of the speed limitations of the sideway-force equipment (such as SCRIM), it has been used for measurements where higher speeds are required, such as motorways and dual carriageways. Tests have been made at at speeds as high as 160 km/h on airfields and 130 km/h became a "standard" high speed for roads.

In this test a small trailer tyre is braked so that the wheel locks for about 2 seconds. The brake torque is measured while the wheel is locked by a system similar to that used for the sideway-force measurement. Results are related to the vertical-force and are expressed as "braking-force coefficients" (BFC)s.

By their very nature tests are of an intermittent rather than continuous nature and for safety reasons require the area under test to be free from other traffic. Therefore the trailer is essentially a research tool for highway purposes and has to give way to indirect methods for routine monitoring purposes, such as a combination of SCRIM and high-speed texture meter.

6.2.3 SCRIM

6.2.3.1 The machine

SCRIM (Sideway-force Coefficient Routine Investigation Machine) evolved from the motorcycle combination testing machines of the 1930s. The first stage of evolution was to install the test equipment into a suitable car. This involved incorporating a fifth wheel, the test wheel, in a front-wheel drive car. The first suitable production cars available were Citroëns and the first test machines were fitted in them, but when a suitable British front-wheel drive car became available in the early 1960s (the Austin 1800) three were converted for SFC measurements.

These cars were only suitable for research purposes as they required a separate vehicle to wet the road surface before test and the processing of the recordings was very time consuming. A special vehicle was therefore designed to overcome these shortcomings and the first of these was built for the TRRL by WDM Ltd of Bristol in 1968.

SCRIM consists of a lorry chassis onto which is mounted a water tank (approximately 3000 litre), together with the test wheel assembly and measuring and recording equipment. The latter computes the average SFC over successive 5, 10 or 20 metre sub-lengths of roads and the results are recorded (originally on punched paper-tape, but now on magnetic tape) together with information on the speed and location of the vehicle.

SCRIM can test a road surface whilst travelling along with traffic and gives a continuous record of SFC. It needs to stop only to refill the water tank or, in the earlier models, to change the paper tape when long routes such as motorways are being tested.

A limitation of SCRIM is its inability to test at speeds greater than 80 km/h. However an estimate of high-speed skidding resistance can be made from knowledge of lower-speed skidding resistance and the macro-texture depth of the surfacing, which can be measured with the high-speed texture meter (HSTM).

The Transport and Road Research Laboratory's Report 737 (Hosking and Woodford, 1976a) gives guidance on the use of SCRIM and Laboratory Report 739 (Hosking and Woodford, 1976c) gives an account of the factors that can affect the measurements. In addition to the factors that affect the slipperiness of the road surface (see 6.2.3.2 below) a number of other factors can affect the measurement itself, an example of which is the speed of test. It is therefore essential to standardise the way in measurements are made before SCRIM (or any other routine skid-testing apparatus) is used for the routine monitoring of road networks and the results related to any target levels. It is also necessary to standardise the terminology used (see 6.2.3.4. below).

6.2.3.2 Factors affecting the slipperiness of the road surface

The factors affecting the slipperiness of a road surface are many. They are discussed in detail in the TRRL's Laboratory Report 738 (Hosking and Woodford, 1976b). They include seasonal variation, temperature, type and composition of the road surfacing material, type and condition of the road, age, traffic density, climate and any contamination of the road surface.

There is a marked seasonal variation in skid-resistance measurements that are obtained on roads in Great Britain (see Fig. 48). Higher values are recorded during cold wet winter months and lower values are recorded during hot dry summer months. A number of explanations of this phenomenon have been put forward, which include the effect of temperature on the resilience of the rubber in the test instrument, differences in the degree of polishing resulting from differences in particle size of the abrasive particles on the road surface, and physical and chemical changes in the binder in the surfacing. It is likely that all these factors contribute to this seasonal variation. In order to overcome the problem of assigning values for such a variable parameter, it is customary to refer to mean summer values which are the mean of at least three (and preferably more) measurements that have been made at well spaced intervals during the period May to September inclusive.

6.2.3.3 Corrections and adjustments

A number of the factors affecting the slipperiness of the road surface and the operation of SCRIM do so in a predictable manner. This allows corrections and/or adjustments to be made to measured values to bring them into line with tests under standard conditions. Factors in this category include the calibration of the equipment, speed, temperature and season. If more than one correction is to be applied the order of correction is calibration, speed, temperature and finally seasonal.

6.2.3.4 Terminology for use with SCRIM data analysis and reporting

Standard terminology has been drawn up by the TRRL in order to overcome possible ambiguities and misunderstandings with regard to the reporting and interpretation of results (Transport and Road Research Laboratory, 1982). This is reproduced below:-

a. Sideway-force coefficient (SFC).

    This is the general term for the ratio of the sideway-force to the normal-force obtained with sideway-force road friction testing equipment. The actual value will depend upon the type of equipment used and the way in which it is used. The term SFC will only be used in this general sense.

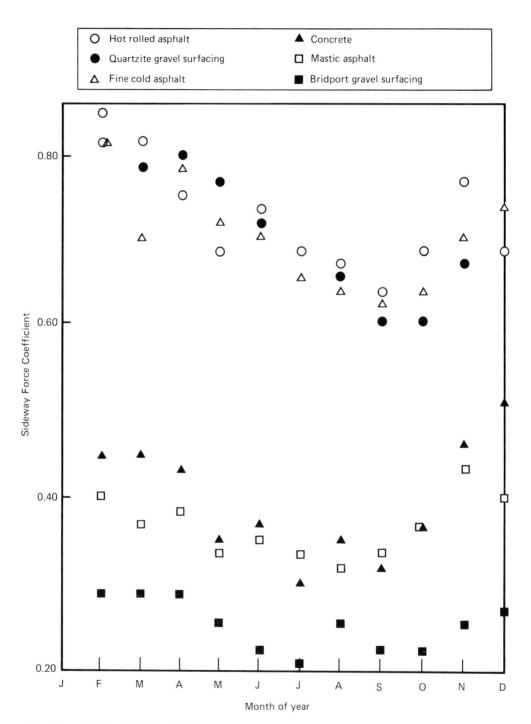

Fig. 48   Variation in SFCs on the TRRL test track during 1966.

b. Raw data.

This is a general term to describe the unprocessed data as a sequence of values output by SCRIM before any analysis or value adjustments have been made.

c. SCRIM reading (SR).

This is the individual frictional measurement as recorded by SCRIM for a single sub-section (5, 10 or 20 metres long according to the recorder setting). It is expressed as a positive, unsigned integer. (Equivalent to the friction ratio x100).

d. Standard testing.

To give immediately usable data a SCRIM must be operated under standard conditions, i.e. with a correctly calibrated recording system, at the correct speed, in the correct track, at a standard temperature and in the correct test season. This is known as standard testing.

e. Corrected.

Clearly it is not always practical to achieve standard testing under normal weather and traffic conditions. Certain factors affect SCRIM measurements in a known way and so corrections may be applied to adjust SCRIM readings to relate more closely to those which would have been obtained in standard testing. When reporting an SR to which corrections have been made, the prefix "corrected" must be applied together with an indication of the factor(s) allowed for, e.g. "Speed-corrected SR".

f. SCRIM coefficient (SC).

A SCRIM coefficient is a SCRIM reading adjusted after any relevant corrections according to the index of SFC (an index explained in the Transport and Road Research Laboratory's Supplementary Report 346 (Hosking and Woodford, 1978)) in force for the SCRIM at the time of the test. A SCRIM coefficient is always expressed as a decimal fraction, to two decimal places.

g. Mean summer SCRIM coefficient (MSSC).

This is the mean of three or more SCRIM coefficients obtained for the same length of road spaced out over the summer testing period (May to September inclusive) of a single year.

h. Equilibrium SCRIM coefficient (ESC).

This is the skid-resistance that a given length of road surfacing reaches after uniform trafficking. It is obtained by taking the mean summer SCRIM coefficients over three or more consecutive years, excluding the first year after opening to traffic.

i. Best estimate.

This prefix must be applied to terms "g." and "h." when incomplete data has been used in

TABLE 20
Average cost (£) of road accidents in Great Britain : 1986

	Fatal injury	Serious only	Slight	All	Damage
Lost output	151,203	1,982	28	3,473	-
Medical & ambulance	1,174	2,069	105	587	-
Police & insurance admin	347	277	208	227	69
Damage to property	1,982	1,568	1,116	1,239	583
Pain, grief and suffering	145,134	12,287	229	5,922	-
Total	299,840	18,182	1,686	11,448	652
Total (revised)*	522,400	18,180	1,690	15,840	658

* A new fatal casualty valuation of £500,000 was proposed by the Minister for Roads and Traffic in 1988. The published revised costs are shown here.

there were 248,000 road accidents in Great Britain in 1986, in which there were 5,382 fatalities, 69,000 serious injuries and 247,000 slight injuries. A total of 321,000 casualties in all.

This Casualty Report also states that the total cost of all road accidents in 1986 was estimated to be £3,800m (revised to £4,890m after re-assessment in 1987). The average costs for each class of accident are detailed in Table 20.

6.3.3 Relation between skid-resistance and accident rate in Great Britain

There have been many attempts to quantify the effect of changes in SFC (and other measures of skid-resistance) on accident frequency. Studies have been complicated by the many other factors that can affect accident frequency and by the relative infrequency of accidents. Most of the successful attempts have taken the form of before-and-after studies, where the change in accident frequency has been related to measurements of skid-resistance before and after re-surfacing with a skid-resistant surfacing. However the most comprehensive study to date was that reported in the TRRL's Research Report 76 (Hosking, 1986). The method used excluded the other possible causes of accidents. Very good correlation was found between changes in SCRIM coefficient (see Fig. 50) and the wet-road skidding rate and also, rather surprisingly, good correlation between SCRIM coefficient and dry-road skidding rate. The most likely explanation of the latter effect was thought to be the reporting of accidents as being on a dry road when, in fact, the road was slightly damp (this is because very slight dampness greatly lowers skid-resistance). It was also thought likely that a surface with a low wet-road skid-resistance would be more slippery under conditions such as when the road is contaminated by oil, dust or other materials.

In Research Report 76 the effect of increasing the skid resistance is quantified in terms of reduction in skidding rate. Skidding rate is a parameter commonly used for assessing skidding accident frequency. It is defined as the number of accidents in which one or more vehicles are reported to have skidded expressed as a percentage of all accidents. In addition to an overall skidding rate there can be separate assessments of wet-road skidding rate and dry-road skidding rate. A conclusion of the study was that an average increase in average SCRIM coefficient of 0.10

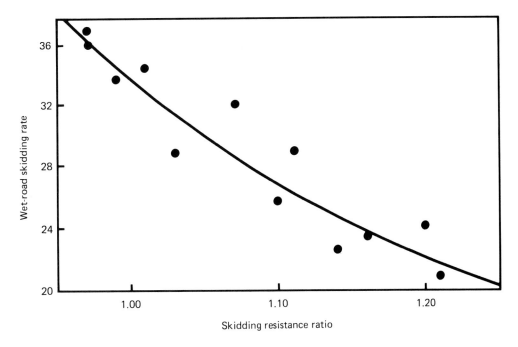

Fig. 50   Relationship between seasonal wet-road skidding rate and skidding resistance ratios.

was accompanied by a reduction in wet-road skidding rate of 13.2. (A reduction of from 35.4 to 22.2).

A further conclusion was that an average increase in average SCRIM coefficient of 0.10 was accompanied by a reduction in dry-road skidding rate of 5.9. (A reduction of from 16.0 to 10.1). Although this reduction was less than the benefit to the wet-road skidding rate, the overall importance is as great because of the larger number of accidents reported on dry roads.

The total number of accidents in Great Britain for 1986 under wet conditions was 83,813 and under dry conditions was 121,452. From these figures it can be estimated that the reduction in number of accidents corresponding to each improvement in SCRIM coefficient of 0.01 would have been 1,106 and 716 respectively for wet and dry roads, giving a total saving of 1,822 personal injury accidents for the country in the year.

A study of the same problem has also been made by Young of Greater London Council (Young, 1985), but was confined to urban classified roads. His figures show an annual saving of 420 accidents on wet roads for an improvement of 0.01 in SCRIM coefficient on urban roads alone.

6.3.4   Some accident/skidding studies made overseas

Examples of some of the studies of the skidding problem that have been conducted in a number of countries are outlined below.

6.3.4.1 The Netherlands

In the Netherlands (State Road Laboratory, 1973) all accidents on state roads in 1965 and 1966 were used in a statistical analysis of the relationship between skid-resistance and accident rate. The accident rate was derived from the number of accidents during a certain period on a selected section of road and the total number of kilometres travelled by vehicles over the section of road during the same period. Friction coefficients for each road section were recorded by the Dutch standard test method (a braking-force coefficient with the test wheel under 84 per cent braking-slip). Relationships were established between friction level and accident rate for both motorways and for other roads under wet conditions.

6.3.4.2 Czechoslovakia

In Czechoslovakia (Zelina, 1973) a statistical analysis covering 24 skid-prone sections of road was made, using the locked-wheel braking force coefficient at 40 km/h to measure skid-resistance. It was found that the percentage of accidents decreased with increasing frictional level of the road surface (see Table 21).

TABLE 21
Friction levels and the percentage of accidents on 24 sections of road in Czechoslovakia

Frictional level (Braking-force coefficient)	< 0.40	0.41-0.50	>0.51
Number of road sections	14	8	2
Percentage of accidents	64	25	11
Mean BFC	0.30	0.44	0.54

6.3.4.3 The Federal Republic of Germany

In the Federal Republic of Germany (Beckmann, 1964) a relationship between skid-resistance and accidents was established in 1964 when, in an investigation covering 32 sections of road, the proportion of accidents that occurred under wet conditions was correlated with locked-wheel braking force coefficients. On most road sections the proportion of accidents in the wet was found to vary between zero and approximately 50 per cent, and averaged about 33 per cent over the whole road network. It was found that sections of road which exceeded this average proportion were likely to have low friction levels.

A later survey (Schulze *et al*, 1974) covering 80 sections, each of 1-8 km in length, of motorways and main roads yielded a relationship between accidents and wet-friction level.

6.3.4.4 Italy

In Italy a particular interest has been taken in British experience with calcined-bauxite/epoxy-resin surface dressings. This led to the issue of the following circular by the Italian Ministry of

Public Works : "At particular spots (bends, intersections, critical points, etc.) carry out special anti-skid treatment with aggregates of high hardness index fixed to the existing surface by means of epoxy resins. This technical measure is especially recommended for city streets, in front of the more dangerous pedestrian crossings, and wherever experience has shown the high incidence of slipperiness to be one of the causes of accidents".

Before-and-after skid-resistance measurements and accident studies demonstrating the effect of the anti-skid treatment are shown in Table 22.

TABLE 22
Comparative figures showing the effect of anti-skid treatments in Italy

Skid resistance (TRRL pendulum)			Average reduction in the total number of accidents during one year
Before	just after treatment	1 year after	
33	88	80	100
46	87	74	50
63	97	-	50
52	96	-	100

6.3.4.5 Japan

In Japan (Ichihara, 1973) about 50 per cent of road traffic accidents were found to occur at intersections. Therefore a survey was made of the effect of anti-skid surface treatments on a number of intersections where a high proportion of the accidents (over 30 per cent) were "head to tail" collisions in wet weather. Skid-resistance measurements were made using the locked-wheel braking force method of the Public Works Research Institute. The increase in frictional levels achieved by the anti-skid treatments are shown in Table 23. Accident figures for two

TABLE 23
Head to tail collisions at 12 intersections before and after skid-resistance treatment

| | Number of head to tail collisions | | | | | |
| | Before | | After | | Change | |
Site	Dry	Wet	Dry	Wet	Dry	Wet
1	2	4	2	1	0	-3
2	7	9	4	1	-3	-8
3	9	15	3	2	-6	-13
4	10	1	1	1	-9	-1
5	1	0	2	0	+1	0
6	4	1	4	0	-1	0
7	1	1	2	0	+1	-1
8	1	4	2	0	+1	-3
9	1	2	3	0	+2	-2
10	6	16	6	0	0	-16
11	6	3	5	1	-1	-2
12	16	13	13	3	-3	-10
Total	64	69	47	9	-17	-60

comparable 4 month periods before and after treatment demonstrated a significant reduction in "head to tail" collisions in wet weather (average 87 per cent) and even in dry weather (average 27 per cent).

6.3.5 The saving in accidents and their cost by improvements in skid-resistance in Great Britain

An estimate has been made of the reduction in road casualties that can be achieved by improving the skid-resistance of our trunk road network (Department of Transport, 1988). Preliminary figures indicated that up to 1800 casualties could be saved each year by the introduction of new requirements for the skidding resistance of trunk roads. Implementation of these requirements was expected to contribute to the Department of Transport's target of cutting road casualty figures overall by one third by the year 2000.

The additional cost of implementation was estimated to be about £9 million a year for the first four years and about £2.5 million a year thereafter. When fully implemented it was estimated that financial benefits from accident reductions on trunk roads would be about £35 million a year, representing a return of £5.50 for every £1 spent.

These standards (see Chapter 7.1.1.4) were introduced early in 1988. Experience after two years (Transport and Road Research Laboratory, 1990) suggests that these preliminary figures may be an under-estimate. At some sites accidents in wet conditions have been reduced by 50 per cent.

6.4 Traffic noise

A review of research (Salt, 1979) indicated that evidence pointed to the interaction between tyre and road surface as being the principal cause of the rolling noise generated by traffic. In the case of lighter road vehicles rolling noise is the main source of total traffic noise and makes a significant contribution to the total noise generated by the heavier vehicles.

In the case of dry roads it is believed that the impact of the tyre on the road is the principal cause of tyre/road noise, this is augmented by the noise produced by radial and tangential oscillations. The relative importance of these three factors is determined by the texture of the road surface. For example a transversely grooved concrete would have a different effect from the random projections in a surface dressing.

Although no general correlation has been found between surface texture and the noise produced, significant correlation has been observed for particular classes of surfacing. Examples are given in Fig. 51 for various bituminous surfacings and concrete surfacings in Great Britain (Salt, 1979). In this paper Salt also reported a finding of great practical value. This was the significant correlation between the noise generated on a surfacing and the loss in skidding resistance when speed is increased from 50 to 130 km/h (Fig. 52). The value lies in enabling test requirements to be formulated which are equitable to both bituminous and concrete surfacing interests, and at the same time strike a balance between road safety and noise.

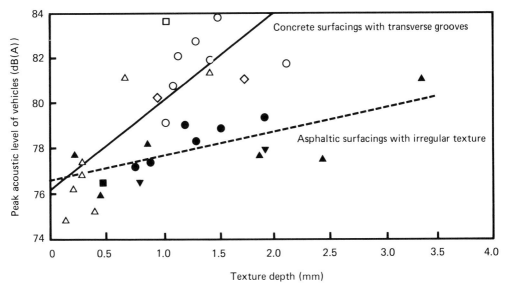

Fig. 51   Relation between the texture depth and the noise of light vehicles at 70 km/h for various bituminous and concrete surfacings.

6.5 Reflecting properties of roads

6.5.1   General

Although the light reflecting properties of road surfacing materials can affect general visibility at night and in tunnels, the most common application of reflecting materials is in road markings. At one time such markings used a white aggregate, such as calcined flint, in a white thermoplastic binder. More recently the better light reflecting properties of ballotini (small glass beads) have led to their replacing white aggregate for most applications. Nevertheless greater skid-resistance is achieved with aggregates such as calcined flint and it is considered advantageous to use them in some circumstances. The more general problem of night visibility is more complex. The reflecting properties of wet (or damp) roads are quite different from those when the road surface is dry, because of the greater specular reflection.

6.5.2   Road surface reflection characteristics

The problems of visibility and glare involve interaction between street lighting and road materials. Considerable interest has been taken in the subject by several countries but no

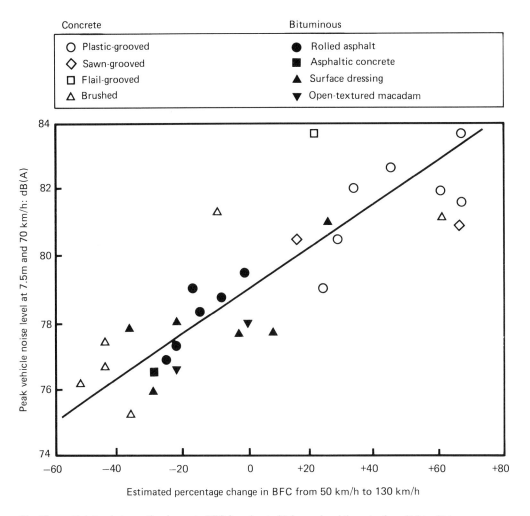

Fig. 52   Relation between the change in BFC from low to high speed and the noise from light vehicles.

generally recognised standards have yet been drawn up to either ensure adequate visibility or limit glare.

The RRL reviewed the problem some years ago (Christie, 1966) and drew attention to the interaction between the preferential reflection from surfaces and their resistance to skidding. Fig. 53 shows the serious fall off in preferential reflection that occurs as resistance to skidding is improved. Realising that there would be a trend towards the general improvement of skid resistance, he concluded that the only feasible method of increasing the luminance of road surfaces under road lighting (and with it the revealing power of the lighting) was by the use of lighter coloured road materials. He also pointed out that such materials would be expensive, difficult to keep clean and could be dazzling in bright sunlight. Concrete surfaces would have an advantage, but bituminous surfacings could be lightened by the use of light coloured aggregates such as calcined flint.

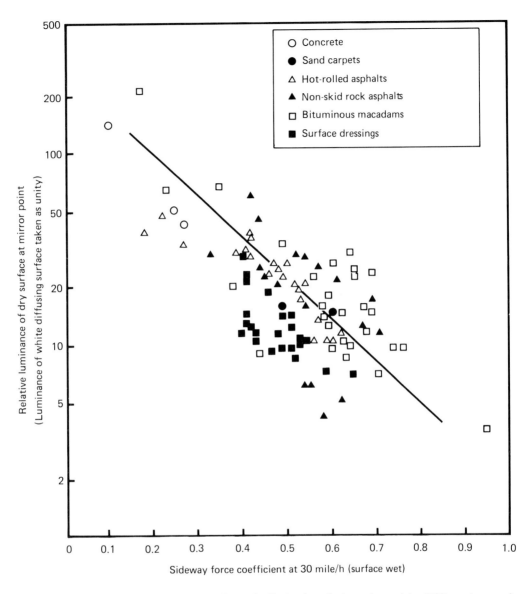

Fig. 53   The relationship between the preferential reflection from the dry surface and the skidding resistance of the wet surface.

An experimental surface of this type was laid on the A217, Garrett Lane, Wandsworth, in 1963 (James, 1965). Here, calcined flint chippings were used in rolled asphalt to lighten the colour of the road surface. It was observed that under street lighting the average road luminance was less than with the former polished surface, but seeing conditions were improved. No change in accident rate was reported.

7 Applications of research and experience

7.1 Resistance to skidding

7.1.1 Skid resistance standards for roads

7.1.1.1 Giles' proposals

Giles proposed the first comprehensive standards of skid-resistance in Great Britain in his paper to the Institution of Civil Engineers in 1956 (Giles, 1957). He divided road sites into four categories: A = "most difficult sites", B = "general requirements", C = "easy sites" and D = "proved sites". The proposed sideway-force coefficient (SFC) requirements being above 0.6, above 0.5, above 0.4 and "no requirement" respectively. These proposals were formulated after comprehensive studies of tyre/road adhesion whilst vehicles were braking and manoeuvring, together with studies at accident black spots. Comparisons of the SFC at accident spots were made with the SFC at similar sites taken at random, and his proposed standards were based on the likelihood of a site being the scene of skidding accidents.

These proposals, shown in Table 24, were intended as a tentative guide and as a basis for further study.

7.1.1.2 Marshall recommendations

Recommendations for skidding resistance were included in the Ministry of Transport's Marshall Committee on Highway Maintenance (Ministry of Transport, 1970) whose report was submitted to the Minister of Transport in 1970. These were based on target values proposed by Giles. The important differences were the omission of the Category D sites, the clearer definition of Category B sites and the reduction of the target for Category A sites from 0.60 to 0.55. These recommendations provided guidance for the maintenance engineer.

7.1.1.3 Transport and Road Research Laboratory's proposals

In 1970 the Transport and Road Research Laboratory set out to try to produce a scheme which would be acceptable from all points of view. This scheme was based on the target values proposed by Giles and the Marshall recommendations, but was improved by incorporating the results of further research. The main points were as follows:

a.   Sideway-force coefficient

The starting point of this study was a consideration of the limitations of the earlier proposals. The recommendations of minimum values of SFC for three or four broadly defined categories of site had the advantage of simplicity but did not fully reflect the fact that a slippery surface is just one of many factors contributing to the incidence of skidding. Two sites fitting into the same category

TABLE 24
Giles' suggested sideway-force coefficients (1957)

Category	Type of site on wet surface	SFC at 30mile/h
A	"Most difficult sites" such as: 1. roundabouts 2. bends with radius less than 500 ft on fast de-restricted roads 3. gradients 1 in 20 or steeper of length greater than 100 yards 4. approaches to traffic lights on de-restricted roads	above 0.6
B	"General requirements" i.e.: roads and conditions not covered by categories A & C	above 0.5
C	"Easy sites" e.g.: mainly straight roads with easy gradients and curves and without junctions and free from any features such as mixed traffic especially liable to create conditions of emergency	above 0.4
D	"Proved sites" e.g.: roads with coefficients below 0.4 which because of factors such as very slow or infrequent traffic cannot be shown by accident studies to be above normal danger	no requirement

can show widely different skidding accident records, because the risk of accidents occurring is influenced by many factors such as traffic speed, road geometry, super-elevation of bends, visibility and many others.

Since the introduction of the earlier proposals a considerable increase in traffic had occurred (59 per cent increase in commercial vehicles between 1956 and 1969) and the polishing effect of the traffic had reached such proportions that it had become impracticable to maintain the target values on many very heavily trafficked lanes of motorways using any naturally occurring aggregate.

The new proposals which were prepared were published in TRRL Report LR 510 (Salt and Szatkowski, 1973). They were based on a site classification system broadly similar to the earlier Marshall recommendation but included a new class of high risk site. The main feature of the scheme was that in place of a rigid system, the minimum value of SFC required on any site should additionally be dependent on a "risk rating" which was to be determined locally by the accident potential of the site. If the mean summer SFC fell below the minimum value the maintenance authority was recommended to initiate remedial action by including the length in the programme for future maintenance work provided the accident record gave no grounds for re-assessment with a lower risk number. The proposals are given in the Table 25.

TABLE 25

Minimum values of skidding resistance for different sites

| Site | Definition | \<td colspan=10\>SFC (at 50 km/h) Risk rating | | | | | | | | | |

Site	Definition	1	2	3	4	5	6	7	8	9	10
A1 (very difficult)	(i) approaches to traffic signals on roads with a speed limit greater than 40 mile/h (64 km/h) (ii) Approach to traffic signals, pedestrian crossings and similar hazards on main urban roads						0.55	0.60	0.65	0.70	0.75
A2 (difficult)	(i) Approaches to major junctions on roads carrying more than 250 commercial vehicles per lane per day (ii) Roundabouts and their approaches (iii) Bends with a radius less than 150 m on roads with a speed limit greater than 40 mile/h (iv) Gradients of 5% or steeper, longer than 100m				0.45	0.50	0.55	0.60	0.65		
B (average)	Generally straight sections of and large radius curves on: (i) Motorways (ii) Trunk and principal roads (iii) Other roads carrying more than 250 commercial vehicles per lane per day	0.30	0.35	0.40	0.45	0.50	0.55				
C (easy)	(i) Generally straight sections of lightly trafficked roads (ii) Other roads where wet accidents are unlikely to be a problem	0.30	0.35	0.40	0.45						

b.   Polished-stone value required

A summarised version of the PSV requirements for use in bituminous surfacings was produced for a range of traffic conditions and included with the recommendations. (See Chapter 7.1.2.2).

c.   Surface texture

The effect of macro-texture on the change in skidding resistance with speed was included in the TRRL proposals to give a comprehensive policy on surfacings.

It was proposed that new surfacings, bituminous and concrete, should have sufficient texture to give the same skidding resistance at high speed as at low and that maintenance intervention should take place when texture was reduced to the point where a 20 per cent drop occurred. Zero drop required 2.0 mm for new bituminous surfacings and 0.8 mm for new concrete surfacings, and 20 per cent drop required maintenance coming into play at 1.0 mm and 0.5 mm respectively. (See Chapter 6.4 "Traffic noise".)

7.1.1.4 Department of Transport's standard

In order to have maximum effect in improving road surfaces any standards need to have mandatory backing. A major deterrent to the adoption of mandatory standards has been the legal consequences of such action.

The first mandatory skidding standards for in-service roads in Great Britain were those published by the Department of Transport in 1988 in their Departmental Standard HD 15/87 (Department of Transport, 1987a) and Advice Note HA 36/87 (Department of Transport, 1987b). These apply to in service trunk roads.

The introduction was made possible following the large scale nationwide National Skidding Resistance Survey (NSRS) undertaken by the Department of Transport; the results of the survey allowed Accident Risk versus Wet Skidding Resistance relationships to be determined for approximately 5000 individual sites covering all the main categories found on the trunk roads. This information, together with earlier TRRL research work, allowed the Department of Transport to select maintainable, cost effective, accident saving investigatory levels for each of the categories. A satisfying aspect of the investigation was how well the main findings of the NSRS investigation confirmed the earlier research work. (Rogers and Gargett, 1991).

Departmental Standard HD 15/87 and Advice Note HA 36/87 require that one third of the trunk road network is tested for skid-resistance each year and the full network over a three year period. Investigatory levels of skid-resistance are prescribed (these are reproduced in Table 26) and, if the skid-resistance is at or below the appropriate level, warning signs must be erected, the maintaining agent notified and the site investigated. If it appears likely that there is an accident problem which is associated with low skid-resistance, a cost and economic appraisal should be undertaken. Sites where surface treatment is required are then short listed in order of priority to enable funds to be allocated to the best effect.

The documents give comprehensive instructions for carrying out skid-resistance surveys using SCRIM and advice on investigatory procedures and on remedial measures.

TABLE 26

Investigatory skidding resistance levels for different categories of site (from Departmental Standard HD/87)

Site Category	Site Definition	Investigatory levels of mean summer sideway force coefficient (at 50 km/h)							
		0.30	0.35	0.40	0.45	0.50	0.55	0.60	0.65
		\multicolumn{8}{c}{Corresponding risk rating}							
		1	2	3	4	5	6	7	8
A	Motorway (mainline)		++++						
B	Dual carriageway (all purpose)-non event sections		++++						
C	Single carriageway - non event sections			++++					
D	Dual carriageway (all purpose)-minor junctions			++++					
E	Single carriageway minor junctions				++++				
F	Approaches to and across major junctions (all limbs)				++++				
G1	Gradient 5% to 10%, longer than 50 m Dual (downhill only) Single (uphill and downhill)				++++				
G2	Gradient steeper than 10%, longer than 50 m Dual (downhill only) Single (uphill and downhill)					++++			
H1	Bend (not subject to 40 mph or lower speed limit) radius <250 m				++++				
J	Approach to roundabout						++++		
K	Approach to traffic signals, pedestrian crossings, railway level crossings or similar						++++		

Site category	Site definition	Investigatory levels of mean summer sideway force coefficient (at 20 km/h)							
		0.40	0.45	0.50	0.55	0.60	0.65	0.70	0.75
		\multicolumn{8}{c}{Corresponding risk rating}							
		1	2	3	4	5	6	7	8
H2	Bend (not subject to 40 mph or lower speed limit) radius <100 m					++++			
L	Roundabout				++++				

++++ Investigatory levels

The usual method of improving the skid-resistance is to re-surface with an aggregate of higher polish-resistance (see 7.1.2 below). The latest document covering this point is the Department of Transport's Technical Memorandum H 16/76 (Department of Transport, 1976). Revision of this Memorandum is being undertaken at the time of writing (1990), but its implementation is difficult because of the need to ensure that the aggregates giving the appropriate resistance to polishing would also provide a surfacing with a sufficiently long life.

7.1.1.5 Local Authorities "standard"

The Local Authorities have published guide-lines on road maintenance in their "Highway Maintenance. A Code of Good Practice" (Local Authority Associations, 1989). The authorities concerned are the Association of County Councils, the Association of District Councils, the Association of Metropolitan Authorities and the Convention of Scottish Local Authorities. It has been called a "Code of Good Practice" since it is not about uniformity in highway maintenance nor is it a document with legal status. Highway authorities will wish to adopt various policies and standards that depend primarily on their assessment of local conditions.

This Code gives information on the monitoring of skid-resistance and on the appropriate action to take if this deteriorates excessively. A table of "investigatory skidding resistance levels for different categories of site" is included that has been based on the Department of Transport's Departmental Standard HD 15/87 (Department of Transport, 1987a). Two additional levels have been added by the Local Authority Associations. One is for category A sites which have been divided in to A1 and A2, with a level of mean summer SCRIM coefficient (MSSC) of 0.30 for the A1 sites where the risk rating is 1. The other is an MSSC of 0.35 for category C sites for an additional risk rating level of 2. (See Table 27).

7.1.1.6 Suggested values for the portable skid-resistance tester

Road Note No. 27 (Transport and Road Research Laboratory, 1969) is now out of print and has been replaced by Messrs. W F Stanley's "Instructions for using the portable skid-resistance tester" (SRT). Road Note No. 27 included a table of "Suggested minimum values of skid resistance" as measured with the portable tester (this is reproduced as Table 28). This table was included to give a general guide as the level of values and was in no way intended to be a requirement. The suggested values were deliberately omitted from Stanley's "Instructions" because it was considered that they could be misleading and might cause problems in relation to the much more satisfactory assessment of skid-resistance that can be obtained with SCRIM.

Skid-resistance values (SRVs) as measured with the portable tester suffer from two major disadvantages as compared with measurements by SCRIM and the BFC trailer: these are small sample area and low test speed. The small sample area (12.6 cm long by 7.6 cm wide) means that a vast number of laborious individual tests must be made in order to obtain an average value for a road length that would be comparable with that obtained by SCRIM (which measures a continuous length of road of tyre width). The problem is particularly great when testing surfacings such as asphalts with 19 mm chippings, where the presence or absence of chippings in the small test area can make a large difference to the skid-resistance, and so moving the instrument by only a few centimetres can greatly change the recorded value.

The problem of test speed is also much greater with the SRT, for example a surface of low

TABLE 27

Investigatory skidding resistance levels for different categories of site (from "A code of good practice")

Site Category	Site definition (Road hierarchy)	\(i\) Investigatory levels of mean summer sideway force coefficient (at 50 km/h)							
		0.30	0.35	0.40	0.45	0.50	0.55	0.60	0.65
		(ii) Corresponding risk rating							
		1	2	3	4	5	6	7	8
A1	Straight (Category 3) & low risk (Category 4)	oooo							
A2	Motorway (mainline) (Category 1)		++++						
B	Dual carriageway (all purpose) - non event sections		++++						
C	Single carriageway - non event sections		oooo	++++					
D	Dual carriageway (all purpose) - minor junctions			++++					
E	Single carriageway - minor junctions				++++				
F	Approaches to and across major junctions (all limbs)				++++				
G1	Gradient 5% to 10%, longer than 50 m Dual (downhill only) Single (uphill and downhill)				++++				
G2	Gradient steeper than 10%, longer than 50 m Dual (downhill only) Single (uphill and downhill)					++++			
H1	Bend (not subject to 40 mph or lower speed limit) radius <250 m				++++				
J	Approach to roundabout						++++		
K	Approach to traffic signals, pedestrian crossings, railway level crossings or similar						++++		

Site category	Site Definition	(i) Investigatory levels of mean summer sideway force coefficient (at 20 km/h)							
		0.40	0.45	0.50	0.55	0.60	0.65	0.70	0.75
		(ii) Corresponding risk rating							
		1	2	3	4	5	6	7	8
H2	Bend (not subject to 40 mph or lower speed limit) radius <100 m				++++				
L	Roundabout			++++					

++++ Investigatory levels (HD 15/87)   oooo Local Authorities Associations additional levels.

TABLE 28
Suggested minimum values of "skid-resistance" (measured with the portable tester)

Category	Type of site	Minimum skid resistance (surface wet)
A	Difficult sites such as: 1. Roundabouts 2. Bends with radius less than 150 m on unrestricted roads 3. Gradients 1 in 20 or steeper of lengths greater than 100 m 4. Approaches to traffic lights on unrestricted roads	65
B	Motorways, trunk and class 1 roads and heavily trafficked roads in urban areas (carrying more than 2000 vehicles per day)	55
C	All other sites	45

macro-texture can yield a high SRV whereas the same surface could give a low value when tested with SCRIM or other traffic-speed testing equipment.

Despite these limitations the SRT can be of considerable value if the limitations are well understood. It can test small areas and in confined sites where tests could not be performed by the speedier equipment. It can monitor the same test area over a period of time if the area is suitably marked, and it has proved invaluable when testing small-scale experimental panels and core inserts. Its greatest value probably lies in the laboratory where it can be used to make tests under carefully controlled conditions, examples are in the polished-stone value, polished-mortar value and polished-paver value tests.

7.1.2 Polish-resistance standards for aggregates

7.1.2.1 Ministry of Transport's Technical Memorandum T2/67

Because of the simpler legal and practical considerations of specifying polished-stone value (PSV) as compared with sideway-force coefficient (SFC), mandatory standards for PSV have appeared before those for SFC. In 1967 the Ministry of Transport set out values in their Technical Memorandum No T2/67. "Polished stone values of aggregate for bituminous wearing courses and surface dressings" (Ministry of Transport, 1967). The limits were based on an examination of the results of full-scale experiments together with data from skidding-accident sites. Three categories of site were specified: "A", "B", and "C", that were similar to Giles' SFC categories. The initial proposals were that Category C sites needed a minimum PSV of 45 in order to maintain the recommended minimum resistance to skidding. Category B and A sites were found to be more exacting, a minimum PSV of 60 being found to be necessary for Category B sites and a PSV of at least 65 for Category A sites. Industry was then consulted and, in the light of a scarcity of high PSV stone, these figures were moderated to 59 and 62 respectively to give a workable specification.

The shortage of the high-PSV stone necessary to allow the implementation of a really satisfactory standard led the Transport and Road Research Laboratory to commence a programme of research aimed at seeking new sources. This is outlined in other chapters of this book.

7.1.2.2 Transport and Road Research Laboratory's LR 510 Proposals

The TRRL's proposals (Salt and Szatkowski, 1973) for a standard for skidding resistance pivoted on the PSV requirement corresponding to the different categories of road site and risk rating. These are given in Table 29.

These proposals called for a considerable usage of high-PSV aggregates. Certain grades of calcined bauxite can achieve values of 75, but were not usually available in chipping sizes. Fortunately calcined-bauxite/epoxy-resin dressing systems can achieve higher SFCs than their PSV indicates because of the small size of the aggregate used. In bituminous surfacings such small sizes cannot be used as the aggregate would become embedded in the binder.

TABLE 29
PSV of aggregate necessary to achieve the required skidding resistance in bituminous surfacings under different traffic conditions

Required mean summer SFC at 50 km/h	PSV of aggregate necessary					
	Traffic in commercial vehicles per lane per day					
	<250	*1000*	*1750*	*2500*	*3250*	*4000*
0.30	30	35	40	45	50	55
0.35	35	40	45	50	55	60
0.40	40	45	50	55	60	65
0.45	45	50	55	60	65	70
0.50	50	55	60	65	70	75
0.55	55	60	65	70	75	
0.60	60	65	70	75		
0.65	65	70	75			
0.70	70	75				
0.75	75					

7.1.2.3 Department of Transport's Technical Memorandum H 16/76

The most current mandatory requirements laid down by the Department of Transport for aggregate properties for new motorway and trunk road construction appear in Technical Memorandum H 16/76 (Department of Transport, 1976). It supplements the 1976 edition of the Department's Specification for Road and Bridge Works.

The Author understands that the TRRL is currently reviewing the PSV/SFC relationship in the light of the more severe present-day traffic conditions. This could lead to a future review of requirements.

The road categories are as given in Table 26. The current requirements for PSV are given in Table 30.

TABLE 30
Minimum polished-stone coefficients for bituminous roads

Site	Traffic in commercial vehicles per lane per day	Minimum PSV	Remarks
A1	Less than 250 cv/lane/day	60	Values include +5 units
	250 to 1000 cv/lane/day	65	for braking/turning
	1000 to 1750 cv/lane/day	70	Risk rating 6
	More than 1750 cv/lane/day	75	
A2	Less than 1750 cv/lane/day	60	Values include +5 units
	1750 to 2500 cv/lane/day	65	for braking/turning
	2500 to 3250 cv/lane/day	70	Risk rating 4
	More than 3250 cv/lane/day	75	
B	Less than 1750 cv/lane/day	55	Risk rating 2
	1750 to 4000 cv/lane/day	60	
	More than 4000 cv/lane/day	65	
C		45	No risk rating applied. Many local aggregates have a PSV well above 45 and normally these should be used

This Memorandum also specifies requirements for the abrasion resistance of the aggregates used. This is because the resistance to skidding is dependent on the durability of the aggregate exposed at the surface. These requirements are given in Table 31.

TABLE 31
Traffic loadings and maximum aggregate abrasion values for flexible surfacings

Traffic in commercial vehicles per lane per day	Under 250*	Up to 1000	Up to 1750	Up to 2500	Up to 3250	Over 3250
Maximum AAV for chippings	14	12	12	10	10	10
Maximum AAV for aggregate in coated macadam wearing courses	16	16	14	14	12	12

*For lightly trafficked roads carrying less than 250 commercial vehicles per lane per day aggregate of higher AAV may be used where experience has shown that satisfactory performance is achieved by aggregate from a particular source.

7.2 Overseas specifications for skid-resistance and aggregates

A survey of the way in which a number of countries were tackling the problems of drawing up skidding standards was made by the Permanent International Association of Road Congresses (PIARC, 1975). The results are outlined in the following paragraphs. At that time some of the proposals referred to were of an experimental nature and there have been some subsequent changes.

7.2.1 Belgium

In Belgium a contractor was required to meet the specification of the Ministry of Public Works concerning skid-resistance for any new construction of state roads. This specification included requirements for aggregate-polishing characteristics and ignition loss of the aggregate as well as the type of chippings or the type of grooving of concrete.

Experimental minimum levels of SFC were stipulated at the opening of a road and during the subsequent period of 3 years. However these SFC requirements were withdrawn in 1979 and replaced by requirements for aggregate and texture.

The following specifications, published by the Belgian Ministry of Public Works, have operated for motorways and trunk roads:

i   Aggregate larger than 8 mm, used in the wearing course, must have a minimum PSV of 50.

ii Texturing is compulsory for concrete surfacings and the texture depth, measured by a depth gauge, is required to be initially between 6 and 10 mm decreasing to not less than 4 mm at the final acceptance.

The original experimental SFC requirements that were later withdrawn were: The minimum value of skid-resistance measured as SFC should be 0.45 at 80 km/h for motorways and roads with four or more traffic lanes and 0.45 at 50 km/h for other roads.

PSV was required to be determined according to the method of BS 812. SFC was measured with the Stradographe fitted with a smooth tyre at an angle of 15 degrees under a load of 250 kg.

7.2.2 Czechoslovakia

National standards have been formulated. These were CSN 7361 77 "Methods of measurements of skid-resistance" and CSN 7361 95 "Criteria for the evaluation of skid-resistance".

7.2.3 Finland

No formal standards were in existence in 1975 but it was usual to include a clause in a contract for new construction specifying that the pavement should not be slippery. It has been observed that the widespread use of studded tyres in Finland tends to roughen the surface of roads in the autumn so that any further restoration of skid-resistance is seldom required.

7.2.4 France

Specifications for skid resistance were indirect in that they defined only materials and methods of construction, because it was considered impracticable to define standards in terms of the actual skid-resistance. It was believed that such standards could not be sufficiently precise to be of real value to the road user and economically viable at the same time. In addition, the inherent variability of measurement due to experimental error, environmental influence and variability of the road itself, was thought to create insuperable problems of interpretation of results against any formal standard. Based on experience of measurements taken with a BFC trailer and the Stradographe, at speed of up to 140 km/h, it was decided to provide a materials-based specification for a skid-resistant surfacing.

In 1969 the Ministere de l'Equipment et du Logement set minimum CPA (coefficient de polissage accelere) requirements for aggregates used in wearing courses of hot-mix asphaltic materials and in surface dressings. Texture depth was specified in terms of the sand-patch method and recommendations of different levels of texture were given for a range of traffic speeds.

CPA (similar to the PSV used in Great Britain) requirements were that values below 35 should not be used in the wearing course. CPAs of 35 to 45 were acceptable for use under favourable site and traffic conditions. 45 to 55 were designated "good", and "very good" aggregates (those with values over 55) were recommended for difficult site and traffic conditions (bends, junctions, high speeds, heavy traffic).

Texture depth recommendations were that smooth surfacings with a depth of less than 0.2 mm should be banned. Fine texture surfacings of between 0.2 mm and 0.4 mm texture depth should be limited to those sites where the speed of motor vehicles only occasionally exceeded 80 km/h as, for example, in urban areas. Medium textured surfacings (those with depths between 0.4 mm and 0.8 mm) were suitable for general traffic speeds between 80 km/h and 120 km/h. Where traffic speeds generally exceeded 120 km/h, coarse textured surfacings (between 0.8 mm and 1.2 mm) were to be used. A final category of "very coarse textured surfacings" were to be used in special cases such as on high risk sites associated with high traffic speeds or on those sites which showed a tendency for ice to form under conditions of high humidity and where temperatures were in the neighbourhood of 0 degrees C.

7.2.5 Italy

In Italy there were no national standards of skid-resistance in 1975. Some authorities, however, included a requirement of roughness such as "not to render the pavement slippery".

Authorities in the Milan area, which includes certain motorways, required the coefficient of friction (as measured by the trailer method of the Instituto Sperimenatale Stradale del TCI) to be not less than 0.55 one year after laying and thereafter not less than 0.45. Motorways must be tested 1 and 5 years after the laying the surfacing.

7.2.6 Japan

In Japan, no mandatory standards existed but two separate schemes were in operation. These were intended to be used as guides both in new construction and in maintenance work. The schemes were:

i   Recommendations by the Public Works Research Unit of the Ministry of Construction were published as "Anti-skidding requirements". These set minimum values of skid-resistance coefficient as measured by the Japanese heavy testing vehicle (a form of BFC measurement) at 60 km/h. These minimum values were reduced by 0.05 when measurements were taken under adverse conditions or on temporary surfaces.

ii The "Anti-skidding requirements" of the Japan Highways Public Corporation (JHPC) demanded that a newly constructed road should have skid-resistance values (measured with the British portable skid-resistance tester) of more than 60. If values less than 55 were found and were confirmed by a repeat measurement, the section was tested with the heavy testing vehicle operated by the JHPC. If the longitudinal BFC was found to be over 0.35 measured at 80 km/h no further action is taken. If it was found to be less than 0.28, the road could not be opened to traffic until remedial steps had been taken to improve the skid-resistance. Where values of 0.28 to 0.35 were found, the road could be opened to traffic but consideration must be given to repair work if justified by subsequent experience.

7.2.7 The Netherlands

Two separate standards existed in the Netherlands in 1975. The first was a minimum value of skid-resistance to be used by highway authorities to establish maintenance criteria. This standard was advisory and not mandatory. The second was a minimum value specified by the road authority in a contract for a new road. The chosen value was legally binding on the contractor who could be penalised for not meeting the specification. In 1975 both of these minimum values were set at a BFC of 0.51, but it was intended that the value for new roads be increased to 0.55.

Standard conditions for the BFC test are laid down, and seven categories of skid-resistance have been formulated in steps of 0.05 units. These range from "dangerous" (less than 0.36) to "very skid-resistant" (greater than 0.70). For individual sites the values are considered in relation to their wet-road accident statistics.

Approximately 4500 km of the State road network was being monitored every year in 1975 with measurements being taken on 3200 sections, each 100 m long. On new roads a 300 m stretch in every 1 km of road was being tested, but if readings were found below the required minimum, the whole 1 km length was checked.

7.2.8 Poland

Recommended values of skid-resistance have been specified in Poland in terms of the locked-wheel BFC using a trailer operated by COBIRTD. A patterned tyre is used under a 400 kg load and at a standard speed of 60 km/h. A coefficient of 0.35 or more signified a good surface. Values of 0.20 to 0.35 were considered satisfactory for most purposes but those below 0.20 indicate a slippery road which should be considered for treatment.

7.2.9 Spain

Mandatory national standards of skid-resistance existed in Spain in 1975, but they were defined only in terms of the polishing characteristics of the aggregate used in the wearing course. The minimum value of the polishing coefficient (determined according to standard NLT 174/69 which is very similar to the British PSV test) required was 40 or 45, depending upon volume of traffic, and in special cases it was being raised to 50 or 55.

Control tests were being carried out by government laboratories. The administration responsible for road construction would not accept aggregates which did not comply with the specification.

7.2.10 Sweden

No formal standards existed in Sweden in 1975. Measurements of skid-resistance were only carried out for research purposes on new surfacings, on special test roads, or on request from local authorities and private companies. These measurements were performed by the Statens Vag-Och Trafikinstitut according to a standard test procedure employing the BV5 Skiddometer. The minimum value of the longitudinal friction-force coefficient at 100 per cent slip (locked-wheel BFC) recommended by the Institute was 0.3 at 80 km/h.

7.2.11 Switzerland

Skid-resistance has been specified in Switzerland in a national standard SNV 640511. Sites were divided into three categories and minimum levels of skid-resistance value (using the British portable skid-resistance tester (SRT)) specified for each category.

For roads with traffic speeds below 80 km/h the target value was 55, the minimum value at the acceptance of contract was 50 and the lowest value to which the road was allowed to deteriorate was 45. Further decrease indicated the necessity of remedial action. For roads with traffic speeds of 80 km/h or more, the corresponding figures were 60, 55 and 50 and for difficult sites (such as bends of radius 150 m, gradients of over 8 per cent, junctions, bridges, exits from tunnels) the values were 65, 60 and 55. National standard SNV 640510 defined the manner in which the SRT should be used, with particular reference to testing for compliance with the specification (SNV 640511) when a completed construction scheme was being accepted from the contractor.

For this purpose at least three sections in each kilometre of road were chosen and the measurements taken in five places, about 10 m apart, at points where the surface was most likely to be subjected to traffic. If the average value for any section was unsatisfactory the number of measurements was increased until the full extent of the faulty area was clearly established. The same pattern of measurement was being used when tests were carried out for inspection purposes.

7.3 Specifications for aggregates

7.3.1 National and International standards

The British Standards Institution (BSI) was founded in 1901 and the first British Standard Specification for aggregates was published in 1913. This was BS 63:1913 "Sizes of Broken Stone and Chippings". Further standards which make reference to aggregates have been added over the years, a list is given in 7.3.2 below. These have been reviewed at intervals and revised or withdrawn as circumstances dictated.

The International Standards Organisation (ISO) was formed in 1947 with the object of setting international standards that would eliminate the differences between the many national standards that had been (and were being) formulated. Some measure of success has been achieved with this laudable object, but progress has been slow. International Standards of particular relevance to the subjects of this book are those dealing with test sieves and with the precision of test methods.

The formation of the European Economic Community (EEC) in 1958 and the European Free Trade Association (EFTA) in 1960 has led to more rapid standardisation between the member countries. These led to the formation of the Committee Europeen de Normalisation (CEN) which has been given the task of harmonising standards within the EEC and EFTA. The target date for the introduction of the CEN standards is 1992. It is envisaged that the first of these will be "framework standards" that make reference to national standards for detailed requirements. These will be superseded in due course by full CEN Standards which all member countries will be obliged to use.

The extra work relating to CEN standards has put ever increasing pressure on the various national standards bodies and the BSI is no exception. When work begins on CEN standards, national standards are no longer revised. Some of the British Standards that are subject to revision at the time of writing may not be produced in a revised form. This means that references in this book to British Standards that are in the course of publication may not always be matched by a published British standard. In this case, reference should be made to the corresponding CEN standard.

Meanwhile a number of international organisations in addition to ISO have been considering harmonisation of standards on a world-wide basis rather than just European. One of the most active has been the Permanent International Association of Road Congresses (PIARC) whose reports to the four-yearly congresses have included recommendations for the harmonisation of the testing of aggregates. This work was carried out in collaboration with RILEM together with an input from ASTM. A summary of their recommendations is included in Chapter 3. Although PIARC has not formulated recommendations for standards of skid resistance, it has summarised the work and "standards" of a number of countries. References to some of these are also made in other chapters.

7.3.2 A chronology of British Standards referring to aggregates

The first British Standard Specification concerned with roadmaking aggregates was BS 63 "Broken stone and chippings" published in August 1913. This and others relating to the use and/or testing of roadstone and aggregates are listed in Table 32. (*Note:* the names of some have subsequently been changed, some have been withdrawn and most have been revised).

TABLE 32
A chronology of British Standards relating to aggregates

Number	Title
BS 63:1913	Sizes of broken stone and chippings.
BS 340:1928	Concrete kerbs, channels and quadrants in Portland cement.
BS 342:1928	Two-coat asphalt (Sand or sand and stone aggregate wearing surface).
BS 343:1928	Two-coat asphalt (Clinker aggregate).
BS 344:1928	Single-coat asphalt (Sand and stone aggregate).
BS 345:1928	Single coat asphalt (Clinker aggregate).
BS 347:1928	Asphalt macadam (Penetration method).
BS 368:1929	Concrete flags in Portland cement.
BS 410:1931	Test sieves.
BS 433:1931	Cold asphalt macadam.
BS 435:1931	Granite and whinstone kerbs, channels, quadrants and setts.
BS 492:1933	Pre-cast concrete partition slabs - solid.
BS 594:1935	Rolled asphalt. Fluxed Lake asphalt.
BS 595:1935	Rolled Asphalt. Fluxed natural asphalt.
BS 596:1945	Mastic asphalt for roads and footways.
BS 598:1936	Methods for the sampling and examination of bituminous road mixtures.
BS 706:1936	Sandstone kerbs, channels, quadrants and setts.
BS 802:1945	Tarmacadam and tar carpets, granite, limestone or slag aggregate.
BS 812:1938	Sampling and testing of mineral aggregates, sands and fillers.
BS 877:1939	Foamed blastfurnace slag for concrete aggregate.
BS 882:1944	Coarse and fine aggregates from natural sources.
BS 1047:1952	Air-cooled blastfurnace slag coarse aggregate for concrete.
BS 1165:1947	Clinker aggregate for plain concrete.
BS 1198:1944	)
BS 1199:1944	) Building sands from natural sources.
BS 1200:1944	)
BS 1201:1944	Concrete aggregate from natural sources.
BS 1241:1945	Tarmacadam and tar carpets: gravel aggregate.
BS 1377:	Methods of test for soil for civil engineering purposes.
BS 1621:1954	Bitumen macadam with crushed rock or slag aggregate.
BS 1690:1950	Fine cold asphalt.
BS 1881:	Methods of testing concrete.
BS 1984:1953	Single-sized gravel aggregate for roads.
BS 2040:1953	Bitumen macadam with gravel aggregate.
BS 4987:1973	Coated macadam for roads and other paved areas.
BS 5328:1981	Methods for specifying concrete, including ready-mixed concrete.

7.3.3 A history of British Standard BS 812

7.3.3.1 Early developments

Before the publication of BS 812 in 1938 the earlier Standard (BS 63:1913) specified certain requirements for broken stone and chippings. Standard gauges for broken stone were rings of 3 inch, 2 1/2 inch, 2 inch, 1 1/2 inch and 1 inch in size. Slots were used to check the greatest lengths by measurement, their sizes being 4 inch, 3 inch, 2 1/2 inch and 2 inch.

An appendix gave details of 12 trade groups of aggregate, comprising : andesite, "artificial", basalt (including basaltic whinstone), flint, gabbro, granite, gritstone, hornfels, limestone, porphyry. quartzite and schist.

7.3.3.2 BS 812:1938 "Sampling and testing of mineral aggregates, sands and fillers"

The preparation of the original edition was undertaken in 1933 and the standard was published in October 1938. It included the following tests:

a.   Part 1. General tests

1. Sieve analysis (grading).
2. Measurement of aggregate shape (percentage flaky and elongated material).

b.   Part 2. Sampling and testing of roadstone

3. Sampling of stone and blastfurnace slag.
4. Apparent specific gravity.
5. Volume weight and voids.
6. Water absorption and density.
7. Dry attrition test. (Deval attrition machine - similar to ASTM Method D2-33.)
8. Wet attrition test.
9. Abrasion test. (Dorry abrasion lap, 1 inch by 1 inch stone cylinders).
10. Impact test. (Page impact machine - 1 inch by 1 inch stone cylinders).
11. Crushing strength. (1 inch by 1 inch stone cylinders).
12. Stability of clinker. (Pat test).
13. Determination of the lime/silica ratio and the sulphur content of blastfurnace slag.

c.   Part 3. Sampling and testing of gravel and sand

14. Sampling of gravel and sand.
15. Density of fine aggregate.
16. Organic impurities in sand. (Sodium hydroxide colorimetric method. Similar to ASTM Method C40-33).
17. Determination of amount of material finer than No. 200 sieve in aggregates. (Decantation method, also a preliminary field decantation test).
18. Water absorption of gravel.
19. Apparent specific gravity. (Same as method 4).
20. Volume weight and voids. (Same as method 5).

d. Part 4.   *Sampling and testing of fillers*

21. Sampling.
22. Density. (Kerosine method).
23. Volume weight and voids.

e. *Appendices*

A. Description and physical characteristics of aggregates. (Includes a 12 trade group classification. Four groups of particle shape. Six groups of surface characteristics).
B. Details of test sieves. (Includes a graphical method of recording results of grading of aggregate).

7.3.3.3 BS 812:1943 "Methods for the sampling and testing of mineral aggregates, sands and fillers

An extended and slightly revised edition was published in March 1943.

The trade groups were reduced to 11 in number by placing andesites in the basalt group, the basalt (including basaltic whinstone) group was simplified to simply basalt group, and the aggregate crushing test appeared as 11B alongside the crushing strength test (11A).

7.3.3.4 BS 812:1951 "Methods for the sampling and testing of mineral aggregates, sands and fillers"

The next edition, published on 25th May 1951, was a complete revision, intended to bring the standard up-to-date and to cover and co-ordinate more closely the test methods for all the various uses of aggregates, sands and fillers. The intention was that reference should be made to this standard for test methods, instead of giving the test methods as appendices to the individual materials specifications, and so promoting uniformity in the methods used in testing aggregates for all purposes. The revision included rationalization of the grouping of the tests.

The main differences between this edition and the earlier one were as follows:

i	Improved detailing of the test methods.
ii	Improved method for determining clay, fine silt and dust (the sedimentation method).
iii	Improved tests for volume weight and voids of fillers.
iv	Improved (thinner) thickness gauge for the flakiness index determination.
v	Omission of the method of determining the water absorption and density of individual rock specimens as they had been found less reliable than tests on aggregate samples. Improvement of the methods for testing samples of aggregates.
vi	Replacement of the Page impact test (on prepared cylinders) by the aggregate impact test.
vii	Retention of the Deval attrition test for 2 inch stone, but omission of the test for 1 inch and 1/2 inch stone because it was considered to give results of doubtful significance.
viii	The aggregate crushing test appeared in this edition, but the crushing strength test was retained.
ix	Tests for slag and clinker (which appeared in the 1943 edition) were omitted as they were considered to be more appropriately included in the specifications for those materials.

x   Estimates of precision (reproducibility) were given in terms of coefficient of variation (standard deviation expressed as a percentage of the mean).

7.3.3.5 BS 812:1960 "Methods for sampling and testing of mineral aggregates, sands and fillers"

The next edition of BS 812 was published on 30th September 1960. The main changes were:

i   A new test for the shape of aggregates by the determination of the angularity number was introduced.
ii The ten-per-cent fines test for resistance to crushing was introduced.
iii The abrasion (Dorry) test was replaced by the aggregate abrasion test.
iv The attrition test was omitted because it had not been found to be useful under the changed traffic conditions.
v   The polished-stone value test was introduced with the object of indicating the extent to which different types of stone will polish under traffic.

7.3.3.6 BS 812:1967 "Methods for sampling and testing of mineral aggregates, sands and fillers"

The next revision, in June 1967, saw the introduction of the following tests:

i   Bulking.
ii Four methods of determining moisture content (oven-drying method, modified drying method, buoyancy method and siphon-can method).
iii The pH method of determining organic impurities. This replaced the earlier caustic soda colorimetric method.
iv The section on sampling was revised and major changes were made to the tests for absolute specific gravity of filler, bulk density and polished-stone value.

7.3.3.7 BS 812:1975 "Methods for sampling and testing of mineral aggregates, sands and fillers"

July 1975 saw the publication of a further revision. The standard was divided in to four parts, each published as a separate volume:

Part I covered sampling, size, shape and petrological classification.
Part II covered physical properties: relative density (formerly termed specific gravity), water absorption, bulk density, voids, bulking and moisture content.
Part III covered mechanical properties: aggregate impact value, aggregate crushing value, ten-per-cent fines value, aggregate abrasion value and polished-stone value.
Part IV was published later (in 1976) and covered chemical properties.

The main feature of this revision was that the methods were converted to metric SI units. The main effects were on sieve sizes (and the resulting single-sized materials) and on the dimensions of the equipment. Attention was drawn to the fact that these changes could have a small but possibly significant effect on test results. The crushing strength test was omitted because of poor precision and its replacement by the aggregate crushing test and ten-per-cent fines test for most contractual purposes.

7.3.3.8 BS 812:1984-1990 "Testing aggregates"

In the 1980s came further revision of which important features were the production of the standard in parts (one part for each test or related group of tests) and the inclusion of precision data. Information on these tests is given in Chapter 4. A list of those expected to be published by 1990 is given in Table 33 (CEN Standards are expected to replace British Standards after this date - see Section 7.3.1 above)

TABLE 33
Parts of BS 812:1984-1990

Part Number	Title
100:1990	Definitions, symbols, common equipment and calibration.
101:1984	Guide to sampling and testing aggregates
102:1984	Methods of sampling
103.1:1985	Methods for determination of particle size distribution. Sieve tests. (Formerly published as Part 103.)
103.2:1989	Methods for determination of particle size distribution. Sedimentation test. (Formerly published as Part 104.)
104:	Petrographical examination.
105.1:1985	Methods for determination of particle shape. Flakiness index.
105.2:	Methods for determination of particle shape. Elongation index.
106:1985	Method for determination of shell content in coarse aggregate.
107:	Methods for determination of particle density and water absorption.
108:	Methods for determination of bulk density, optimum moisture content, voids and bulking.
109:	Methods for determination of moisture content.
110:	Methods for determination of the aggregate crushing value.
111:	Methods for determination of the ten percent fines value.
112:	Methods for determination of the aggregate impact value.
113:	Method for determination of the aggregate abrasion value.
114:1990	Method for determination of the polished-stone value.
115:*	———
116:*	———
117:1988	Method for determination of water-soluble chloride salts.
118:1988	Method for determination of the sulphate content.
119:1985	Method for determination of acid-soluble material in fine aggregate.
120:1989	Methods for determination of drying shrinkage.
121:	Soundness.
122:*	Methods of estimating the contents of deleterious materials in sands.
123:*	Method for the petrographical examination of concrete aggregates for the evaluation of alkali-reactivity potential.
124:1989	Method for determination of frost-heave.

*Part 115 has been withdrawn, 116 is not allocated and 122 and 123 will probably be incorporated in the relevant CEN standard.

7.4 Current British Standards for aggregates and roadstone

A list of current British Standards for aggregates and roadstone is given in Table 34.

TABLE 34
Current British Standards for aggregates and roadstone

Number	Title
BS 63:Part 1:1987	Specification for single-size aggregate for general purposes.
BS 63:Part 2:1987	Specification for single-size aggregate for surface dressing.
BS 340:1979	Specification for pre-cast concrete kerbs, channels, edgings and quadrants.
BS 368:1971	Specification for pre-cast concrete flags.
BS 410:1986	Specification for test sieving.
BS 435:1975	Specification for dressed natural stone kerbs, channels, quadrants and setts.
BS 594:1985	Specification for hot rolled asphalt for roads and other paved areas.
	Part 1 Constituent materials and asphalt mixes.
BS 598:1974	Sampling and examination of bituminous mixtures for roads and other paved areas.
	Part 2 Methods of analytical testing.
BS 812:1984-1990	Testing aggregates (See Table 33).
BS 882:1983	Specification for aggregates from natural sources for concrete.
BS 1047:1983	Air-cooled blastfurnace slag for use in construction.
BS 1165:1985	Clinker and furnace-bottom ash aggregate for concrete.
BS 1198,1199 & 1200	Re-numbered BS 1199 and 1200.
BS 1199,1200:1976	Specification for building sands from natural sources.
BS 1438:1983	Specification for media for biological percolating filters.
BS 1690:1967	Cold asphalt.
BS 1796:1985	Methods of test sieving.
BS 1881:1970	Withdrawn and replaced by many parts.
	Part 103:1983 Determination of compacting factor.
	Part 116:1983 Compressive strength of concrete cubes.
	Part 117:1983 Tensile splitting strength.
	Part 118:1983 Flexural strength.
BS 1984:1967	Withdrawn and replaced by BS 63.
BS 2787:1956	Glossary (Replaced by BS 6100).
BS 3681:1973	Methods of sampling and testing lightweight aggregates for concrete.
BS 3797:Part 2:1982	Specification for lightweight aggregates for concrete (metric units).
BS 4328:1981	Methods for specifying concrete, including ready mixed concrete.
BS 4987:1988	Coated macadam for roads and other paved areas.
	Part 1 Specification for constituent materials and for mixes.
BS 5309:1976	Methods for sampling chemical products.
	Part 4 Sampling of solids.
BS 5328:1981	Methods for specifying concrete including ready-mixed concrete.
BS 5497:1987	Precision test methods
	Part 1 Guide for the determination of repeatability and reproducibility for a standard test method by inter-laboratory tests.
BS 5835:Part 1:1980	Compactability test for graded aggregates.
BS 6100:6.3:1984	Glossary - Aggregates. (Replaced BS 2787:1956).
BS 6543:1985	Industrial by-products and waste materials.

7.5 ASTM Standards for aggregates

A list of current ASTM Standards for aggregates is given in Table 35.

TABLE 35
ASTM Standards for aggregates

Part Number	Title
ASTM C29-78	Test method for unit weight and voids in aggregate.
ASTM C30-73	Test method for voids in aggregate for concrete.(Discontinued, replaced by C29).
ASTM C33-86	Specification for concrete aggregates.
ASTM C40-84	Test method for organic impurities in fine aggregates for concrete.
ASTM C70-88	Test method for surface moisture in fine aggregate.
ASTM C87-83	Test method for effect of organic impurities in fine aggregate on strength of mortar.
ASTM C88-83	Test method for soundness of aggregates by use of sodium sulfate or magnesium sulfate.
ASTM C117-87	Test method for materials finer than 75 µm (No.200) sieve on mineral aggregates by washing.
ASTM C123-83	Test method for lightweight pieces in aggregate.
ASTM C127-88	Test method for specific gravity and absorption of coarse aggregate.
ASTM C128-88	Test method for specific gravity and absorption of fine aggregate.
ASTM C131-87	Test method for resistance to degradation of small size coarse aggregate by abrasion and impact in the Los Angeles machine.
ASTM C136-84	Test method for sieve analysis of fine and coarse aggregates.
ASTM C142-	Test method for clay lumps and friable particles in aggregates.
ASTM C227-87	Test method for potential alkali-reactivity of cement-aggregate combinations (mortar bar method).
ASTM C235-68	Scratch hardness of coarse aggregate particles. (Discontinued).
ASTM C289-87	Test method for potential reactivity of aggregates (chemical method).
ASTM C294-85	Descriptive nomenclature of constituents of natural mineral aggregates.
ASTM C295-85	Practice for petrographic examination of aggregates for concrete.
ASTM C342-85	Test method for potential volume changes of cement-aggregate combinations.
ASTM C535-89	Test method for resistance to degradation of large-size coarse aggregate by abrasion and impact in the Los Angeles machine.
ASTM C566-84	Test method for total moisture content of aggregate by drying.
ASTM C586-86	Test method for potential alkali-reactivity of carbonate rocks for concrete aggregates (rock cylinder method).
ASTM C641-86	Test method for staining materials in lightweight concrete aggregates.
ASTM C682-87	Recommended practice for evaluation of frost resistance of coarse aggregates in air-entrained concrete by critical dilation procedures.
ASTM C702-87	Method of reducing field samples of aggregate for testing size.
ASTM D2-33	Test for abrasion of rock by use of the Deval machine. (Discontinued 1972.)
ASTM D75-87	Practices for sampling aggregates.
ASTM D289-63	Abrasion of graded coarse aggregate by use of the Deval machine. (Discontinued 1971, replaced by C131 and C535).
ASTM D1411-82	Test method for water soluble chlorides present as admixes in graded aggregate road mixes.
ASTM D1664-85	Test method for coating and resistance to stripping of bituminous-aggregate mixtures containing adhesion promoting agents.
ASTM D2419-79	Test method for sand equivalent value of soils and fine aggregate.
ASTM D3042-86	Test method for insoluble residue in carbonate aggregates.
ASTM D3319-83	Recommended practice for accelerated polishing of aggregates using the British wheel.
ASTM D3398-87	Test method for index of aggregate particle shape and texture.

8 Author's thoughts

The Authors of this series of books have been requested to provide a chapter of their own thoughts. Already some views have been interwoven in the main text, but further points are made below.

8.1 Sources and production of road aggregates

8.1.1 General

In Great Britain, solid rock suitable for the manufacture of road aggregates is almost exclusively quarried from the earlier rock formations; this means that there are few quarries to the east of a line drawn between Lyme Regis and The Wash. This deficiency has, to some extent, been made good by the occurrence of flint gravels in this eastern area, but gravel pits have a shorter life than quarries and so affect larger areas of land.

With few exceptions aggregates need to be of low cost. Exceptions include the special polish-resistant aggregates, which can be highly cost effective both in in terms of durability and reducing accidents. Low cost means not only of low cost in extraction and processing, but also in cost of transportation to the road site. The transportation cost is often many times the cost of the aggregate ex plant. The early development of the industry led to the opening of a large number of quarries, gravel pits and slag processing plants scattered around our country. In addition, it was customary to extract aggregates and fill material from convenient "borrow pits" in areas where suitable material was available. However environmental and commercial considerations have led to more restrictive development of sources.

The current trend to produce materials from a few large sources has also applied to aggregate production to some extent; consequently the number of working quarries has steadily reduced during this century. Nevertheless the relatively high cost of transport has tended to put a brake on this process. With pressures on land usage and environmental considerations, aggregate extraction industries are subject to constraints. Matters have been made even more difficult by the recent large road and housing construction programmes. This has led to consideration of concentrating production to a few large sources ("super quarries") and the development of previously untapped sources. In addition greater attention has been turned to the greater use of industrial wastes, which would yield a double benefit with reduction in the need to dispose of these materials, and to the recycling of road materials.

8.1.2 Super-quarries

There are already several examples of "big being beautiful". Some of these "super-quarries" are geared to the distribution of aggregates by rail; examples are Foster Yeoman's Torr Quarry in Somerset with an annual production of some 5 million tonnes of limestone, and Redland Roadstones's comparable production of pink granite from Mountsorrel Quarry in Leicestershire. These producers use special railway trains to minimise transport costs when distributing the aggregate to strategically placed depots.

Other quarry owners have sought wider horizons and ship their aggregate by sea. An example was Penlee Quarry near Penzance in Cornwall; the hornfels produced there is probably the strongest stone in Europe with a crushing strength of over 100,000 lb/sq. in. (700,000 kN/sq.m.). It was shipped to the Thames and widely used in the London area. Other coastal quarries have operated in Wales, Scotland and the Channel Islands.

Giant quarries using sea transport are now coming into production. An example is Foster Yeoman's Glendsanda Quarry in the Highland Region of Scotland. This uses large bulk cargo ships to provide a very inexpensive form of transport compared with road and rail. It has the additional advantage of being able to supply the many coastal areas that are deficient in readily available aggregates. It is planned to produce as much as 7.5 million tonnes of aggregate a year from this source, and to supply aggregate depots in London, Holland, Belgium, Germany, the Eastern States of the USA, the Caribbean, the Gulf of Mexico, the Mediterranean and Africa.

Although the localisation of aggregate production to a few large quarries such as these has considerable environmental advantages, there are other forces that tend to maintain a greater diversity of sources. Such diversity has the advantage of being able to provide low cost aggregates to customers near these quarries. It also allows for the specialist needs, such as polish-resistance. The cost of production of such aggregates is high because not only does their abrasive nature lead to heavy wear on processing plant, but also because they are usually interbedded with considerable quantities of low-grade or waste material such as shale and soft sandstone. The specifications for such aggregates are also more severe than those for aggregates for other purposes, particularly with regard to grading and shape, and so call for a specialist processing plant. The economics of such production are such that the producers need to maintain a premium price for a premium product but, as such stone has no advantage over other more easily produced aggregates for purposes other than polish-resistance, such sources are unlikely to be economically viable on the super-quarry scale.

8.1.3 Mining

High cost of extraction is the major deterrent to extraction by mining. Apart from colliery shale, which is not deliberately extracted, only two examples in Great Britain are known to the Author. One is a limestone aggregate mined in Sussex from the floor of a gypsum mine. This aggregate is competitively priced because of the shortage of hard stone in this area and because the mine had already been created for other purposes. Mine haulage in this instance is less costly than usual because of the shallow nature of the shaft which allows removal by normal quarrying methods. The second example, also a limestone, is that mined at Middleton in Derbyshire.

Nevertheless if environmental pressures become extreme, mining could provide an alternative to current methods of aggregate extraction, albeit at a price. The Verney report (Verney, 1976) includes a section on mined aggregates.

8.2 General requirements of roadmaking aggregates

It is not envisaged that there will be any immediate change in the general requirements for aggregates used in road engineering.

In the longer term there will be a decline in construction with bituminous materials as the world's oil resources become depleted. This might be self-compensating as it could lead to a reduction in road traffic. A more likely alternative is that the shortfall would be met by the wider use of concrete. This in turn is likely to lead to increased interest in the various methods of providing a better skid-resistant finish to concrete surfacings. These surfacings have yet to be seriously studied in Great Britain, but have already been the subject of considerable research particularly in Belgium and Canada.

8.3 Tests for roadstone and aggregates

8.3.1 General

The choice of test procedures is being made in the CEN Standards, which will replace the various national standards when the open market is introduced in the European Community in 1993.

Many of the test procedures being used in different countries have resulted from a similar evolutionary process. First, a researcher develops a test as a research tool. This test procedure is then studied by other workers in other countries and in the process would be modified. If the test proves satisfactory then it would be adopted as a national standard and in this process it would probably undergo further change. More changes are possible at every subsequent revision of the standard. Such evolutionary processes have led to a variety of test procedures in different countries all stemming from the same origin. An example is the British Standard modified ten-per-cent fines test which started in Germany as a simple cylinder crushing test in which the sample of dried aggregate was crushed under a standard load in a cylinder and the degree of crushing assessed by the change in fineness modulus, and ended with the results being expressed as the load required to produce a specified percentage of fine material.

The influence of the ISO, PIARC, RILEM and, more recently, CEN has not only tended to halt this trend, but has already started to reverse it. Hopefully we shall be left with better specifications, improved by the experience of the various countries concerned.

8.3.2　Grading of aggregates

The cumulative curve of aggregate passing a series of test sieve sizes is probably the most satisfactory method of describing the grading of an aggregate for most purposes. This is usually implemented in specifications as a table of cumulative percentages of aggregate passing a chosen set of test sieves.

However this method has been found to be insufficient for assessing "single-sized" chippings for surface dressings and hot-rolled asphalt. For these applications the "single-sizeness" is considered to be of particular importance. This need has been met in standards such as BS 63 by specifying a minimum requirement for the percentage of the "required size" in addition to a range of percentages of aggregate passing selected test sieves (examples are given in Tables 9 and 10).

Although this method of specification has proved to be reasonably satisfactory the Author has given consideration to an alternative that more closely approaches the ideal. This is to specify by mean size and standard deviation. These two values would give a measure of both the actual size and the degree of scatter.

At first sight this measurement would seem to involve much tedious work, but early research on aggregates at the Road Research Laboratory offers a simple solution to the problem. This work was concerned with the distribution of size of aggregate within the "single-sizes". Each of the single-sizes (e.g. 3/4 to 1/2 inch chippings - nowadays it would have been 20 mm to 14 mm) was subdivided into five parts, and the size distribution plotted. This work showed that, apart from the very fine material, the distribution of "single size" aggregates closely approximated to a statistical normal distribution. This meant that it yielded a straight line when the cumulative mass passing each sieve was plotted on an arithmetic/probability scale. It also meant that a "single size" aggregate could be easily and effectively specified in terms of mean size and standard deviation, because separation would only be required on 3 or 4 sieves in order to construct the straight line. The mean could then be read off from the 50 per cent ordinate and the slope of the line would give the standard deviation.

Another possibility for the improved measurement of the particle size of aggregates lies in the use of optical methods. Such methods have been developed for sizing particles for a number of industrial purposes. Most have been concerned with microscopic particles, but it should be possible to apply the same principles to the measurement of particles of aggregate. Such a method could yield a rapid and accurate method of monitoring the grading of aggregates. It would also lend itself to the "mean and standard deviation" approach to assessing single-sizeness.

8.3.3　Razor-blade shape

Flaky particles can be detrimental to the performance of mixtures, and various shape measurements have been developed and limits written into specifications. However the amount of flaky material is normally assessed by mass rather than by number, because of the tedious nature of manual counting. It seems obvious that it is the number of such particles that has a greater effect on mixtures rather than their mass: the same mass of razor-blade-like particles affects mixes far more than the same mass of slab-like particles that just fail to meet the limiting thickness. The less demanding flakiness index requirement for gravel, compared with that for crushed rock, that is specified in BS 882 (Natural aggregates for concrete) was introduced to allow the slabby Jurassic limestone gravels that occur in some parts of the country to be used as a concrete aggregate.

It is possible that optical methods might be developed to take the tedium out of manual counting and provide a test result that more satisfactorily reflected the behaviour of the aggregate, indeed optical methods might be developed to measure the shape itself.

8.3.4 Classification

It seems to be part of human nature to endeavour to classify all the phenomena that we encounter. The classification of aggregates is no exception and there have been many attempts to provide a working system based upon their geology. Some of these attempts have had the object of trying to assess the value of an aggregate by classifying by petrographical composition so as to reduce, or even obviate, the need for engineering tests. Others have taken the simpler approach of dividing aggregates into a number of broad geological groups.

None of these attempts has as yet proved particularly successful. This is not unexpected for a product derived from a diverse range of basic materials. It would be better to concentrate on a classification based upon engineering properties.

8.4 Special requirements of aggregates for skid-resistance

The finding of large resources of highly polish-resistant gritstone has tended to reduce the need for general-purpose polish-resistant synthetic aggregates. Nevertheless there is still a considerable demand for aggregates for special resin-bound skid-resistant systems. The aggregates used for this purpose are small in size (3 mm nominal size) and need to be exceptionally hard and durable. Corundum (alpha-alumina) based aggregates appear to be the most economic for this purpose and calcined bauxites have been the only materials extensively used. At present all this type of material is being imported into the United Kingdom and it is both scarce and expensive.

Attention has already been drawn to the possibility of calcining UK deposits of bauxite, such as those occurring in Antrim in Northern Ireland, and results have been promising. Further investigation of this possibility is desirable now that it is becoming increasingly difficult to maintain adequate skid-resistance on some of our heavily trafficked road sites and accident black-spots.

The considerable saving in accidents and their cost by the implementation of new skidding standards on trunk roads by the Department of Transport (see Chapter 6.3.5) is likely to stimulate further interest in the possible manufacture of polish-resistant synthetic aggregates. Such aggregates are likely to be highly accident/cost effective on sites where it is difficult to maintain a high level of skid-resistance with conventional aggregates.

8.5 Road surface characteristics

8.5.1 The effect of tyre compound and tread pattern on the significance of skid-testing with SCRIM and HSTM

The value of the measurement of macro-texture in assessing the benefits of low-resilience tyres is briefly referred to in Chapter 6.1.1. The Author's views of the significance of tyre hysteresis and tread pattern on routine skid-testing are set out in more detail in the following paragraphs.

8.5.1.1 Rubber properties in relation to texture and speed

Studies (Sabey *et al*, 1970) were made to put in perspective the relative roles of the drainage and energy loss (hysteresis) properties of tyres in providing good adhesion to the road. Important findings were as follows:

i   Skidding is caused by the lubricating effect of relatively thin water films, not aquaplaning with thick films.
ii   Changes in frictional performance with speed of sliding are different for different rubber compounds to the extent that their order of merit can change.
iii   The behaviour of the test tyres (of natural rubber) used on skid test machines such as SCRIM may not be matched by the behaviour of tyres made with other rubbers.

8.5.1.2 High-hysteresis rubber compounds

It has been found that the hysteresis properties of tyre tread rubber can make a major contribution to improving tyre/road adhesion (Tabor *et al*, 1958; Sabey *et al*, 1964). High-hysteresis rubber (low-resilience rubber) has been variously termed "dead" rubber and "cling" rubber because of its better adhesion properties.

To exploit the potential benefits of high-hysteresis tread materials, the road must have sufficiently large and angular projections in its surface to deform the rubber of the tyre tread, even in the presence of a water film (Lupton, 1968). Substantial improvements in adhesion are obtained with the use of high-hysteresis tyres on surfaces with a rough macro-texture, while the same tyres show little improvement on smoother surfaces. Fig. 45 shows examples of the effect of tread materials with rubber of high- and normal-hysteresis properties on smooth and rough macro-texture surfaces respectively.

8.5.1.3 Tyre pattern

Research has also demonstrated (e.g. Lupton, 1968) the effect of smooth and rough macro-texture on "bald" and patterned tyres. The results (Fig. 54) show that on a smooth surface much lower coefficients of friction are achieved with a smooth tyre than with one with a patterned tread. Conversely this figure also shows little difference in performance between the two tyres on surfaces with a rough macro-texture.

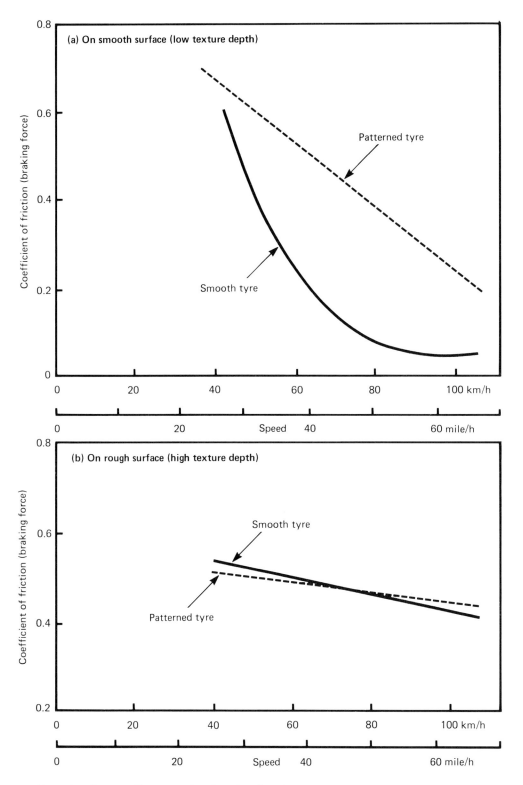

Fig. 54   Tread pattern effect on smooth and rough surfaces (wet)

8.5.1.4 Routine skid-resistance testing

The demands for routine skid-resistance testing machines have led to the development of SCRIM and texture meters. The SCRIM machine has been designed to measure large road networks. In order to obtain meaningful results it has been necessary to standardise the testing parameters such as speed and the characteristics of the test tyre.

The speed is normally standardised to 50 km/h but tests are made at other speeds under certain circumstances, such as when testing tight bends or high-speed roads.

The test tyre tread material has been standardised to a natural rubber composition that has been used for all SFC testing in Great Britain. This has been done not merely to ensure continuity of results but also because high-hysteresis rubber compositions would have a very short test life. Also the results obtained would not apply to the heavier road vehicles that use natural rubber compounds, because of the greater stressing of their tyres.

Another feature of the test tyre is its smooth tread "pattern". This was chosen so that the progressive wear of the tyre would not influence results and because it gave results that would relate to the worst tyre condition.

8.5.1.5 Discussion

The factors outlined above need to be taken into account when considering the significance of present-day methods of monitoring the skid-resistance of roads in Great Britain. Because vehicles are now legally required to maintain a depth of pattern on their tread, the need for a rough macro-texture on lower-speed roads may appear to be reduced. However the general use of high-hysteresis rubber tyres on most light vehicles has created a new need for adequate macro-texture depth for all roads. The significance of these factors in relation to routine monitoring is examined in the following sub-sections.

a.   SCRIM

SCRIM uses a test tyre with a smooth tread and a normal-hysteresis rubber, whereas most present-day vehicles use tyres with a patterned tread and a substantial proportion are of high-hysteresis rubber. These conditions lead to the following:

i   Compared with vehicles using high-hysteresis rubber and patterned tyres (i.e. most lighter vehicles), SCRIM will under-estimate the advantages of rough macro-textured roads because of the hysteresis effect and over-estimate the disadvantages of smooth macro-textured roads because of the tread effect.

ii   Compared with vehicles using normal-hysteresis rubber and patterned tyres (i.e. heavier vehicles), SCRIM will correctly estimate the skid-resistant properties of rough macro-textured roads because of the use of the same rubber and over-estimate the disadvantages of smooth macro-textured roads because of the tread effect.

The overall result is that SCRIM will tend to "fail safe" with respect to the assessment of the skid-resistance of roads. However it will fail to recognise the benefits that roads of rough macro-texture will achieve with high-hysteresis tyres.

b.   Texture measurement

The high speed texture meter (HSTM) and other instruments that give a direct measure of surface macro-texture can be expected to provide a satisfactory estimate of hysteresis benefits. However indirect measurements of texture such as are employed in outflow meters would not correctly assess these benefits on porous surfacings.

8.5.1.6 Conclusions

The main conclusions that the Author has drawn from these mainly theoretical considerations are as follows:

i   In considering the skid-resistance of roads, the present tendency to view texture requirements as simply a drainage problem on high-speed roads is not enough: more attention should be paid to the needs of modern tyres.

ii Ensuring a rough macro-texture on all roads could lead to a greater fulfilment of the benefits of modern low-resilience tyres.

iii The test tyre used in making SCRIM measurements differs not only in hysteresis characteristics from the tyres used by most lighter vehicles, but also in using a smooth tyre as compared with the patterned tyres used on all vehicles. However road assessment with SCRIM will normally "fail safe" when related to actual tyre performance, and the figures obtained have been found to correlate well with accident data.

iv Monitoring all categories of roads with a suitable texture meter would identify sites where road safety might be improved by increasing macro-texture, and which would not be identified by SCRIM alone.

v   Texture measuring equipment of the outflow meter types would not be as suitable as equipment giving a more direct assessment of macro-texture.

8.5.2   Skidding on dry roads

The skidding characteristics of dry roads have not been studied to the same extent as wet roads, because their coefficient of friction is usually high. However it has been shown (Hosking, 1986) that frequency of skidding accidents reported on dry roads shows significant correlation with wet-road skidding resistance as measured by SCRIM. This was attributed to reporting "errors" (a slightly damp road greatly lowers skidding resistance but is often reported as being "dry") and the likelihood that a low-SFC road would be more slippery when contaminated with material other than water (oil, dust, etc.).

Because the benefits of high-hysteresis rubber tyres are also influenced by the macro-texture under dry road conditions, the findings of a study made (Knapp, circa 1975) to correlate vehicle speed and length of skid mark under dry road conditions for the purposes of traffic accident reconstruction has been examined.

A main feature of the study was that two distinct relationships were found between the stopping distance and speed for private cars and heavier vehicles respectively. A linear relationship between the square of the speed and the stopping distance was obtained for each class of vehicle and the calculated coefficient of friction was found to be 0.83 and 0.64 respectively. Several possible contributory factors to this difference were suggested, these included a possible difference in tyre composition.

Although "not proven" it seems likely that the difference in tyre rubber was a major factor influencing these relationships. It is noteworthy that the frictional advantage of the lighter vehicles was similar (0.19 in coefficient of friction) to the difference found between normal- and high-hysteresis rubber on wet roads (Lupton, 1968).

Knapp's report included an anomalous set of results for a car at a road site where the calculated coefficient of friction was only 0.67. He suggested that unusual surface conditions may have been the cause. Although inadequate texture depth may have been the cause, the use of a natural rubber compound in the tyres would be a valid alternative explanation.

It is concluded that any benefits of a rough macro-texture with respect to high-hysteresis tyres are likely to extend to dry road conditions.

8.5.3 Further thoughts about texture requirements

In the above paragraphs (8.5.1 and 8.5.2), the Author has expressed his thoughts about the possible road safety benefits that might be achieved by increasing the macro-texture of a wider range of roads. Although his conclusions are based on theoretical considerations backed up by limited experimental evidence, he feels that there is a strong case for the need of additional research in this area.

There are also other factors that would need to be studied. The cost of surfacing a substantial proportion of roads with materials giving a greater texture depth would be higher and there would be a number of possible disbenefits to consider. These would include the higher cost of monitoring texture, the higher cost of maintenance, additional traffic disruption resulting from more frequent maintenance, greater rolling noise, increased rolling resistance and greater tyre wear.

8.5.4 The effect of present-day trafficking on the PSV/SFC relationship

Increases in traffic levels, advances in tyre technology and vehicle performance are placing ever increasing demands on road surfacings. Existing "standards" are based on research findings established twenty years ago (Szatkowski and Hosking, 1972). This work did not encompass roads bearing the very high traffic flows that we find to-day. It also lacked data to allow a full study of lower traffic levels. The PSV/SFC relationship needs further study under present-day conditions, including the effect of traffic weights as well as numbers. Environmental pressures are also leading to the need for materials that minimise noise emissions from roads. However, these materials may not provide adequate levels of skid resistance. Research is needed to explore the possibilities of developing new surfacing materials such as pervious macadams to provide an optimum solution to the problems.

8.6 Applications of research and experience

8.6.1 Specification limits

8.6.1.1 Limiting values

In reply to a question on specification limits asked at a lecture-course in the 1950s, "Fred" Shergold pointed out that a severe limit could be set that would ensure that all aggregates used would be satisfactory, but it would, at the same time, exclude a number of satisfactory aggregates. Conversely, a less severe limit would allow all satisfactory aggregates to be used but would not exclude some unsatisfactory ones. Even though there have been many studies of aggregates and improvement in the test procedures since this pronouncement, Shergold's reply is still a valid one.

The reason for this situation is that aggregates are manufactured from a very wide range of materials and although good correlation is found between tests and performance for individual classes of aggregates, there is often a different relationship for different classes. An example of this is the polish-resistance of blastfurnace slags, which has been found to be better than most other aggregates of the same polished-stone value.

8.6.1.2 Local knowledge

The requirements for an aggregate for use in highway engineering vary considerably with the type of construction, the layer in the construction, the traffic conditions and the climatic conditions. Over and above these basic needs, engineers have tended to add a generous "factor of safety" to allow for variation between consignments of aggregates and for imprecise knowledge of the minimum requirements. Conversely many engineers have experience of the satisfactory nature of local products that are excluded from general specifications, and would like to make use of them. Certain metallurgical slags come into this category.

8.6.1.3 Methods of specification

In the past, particularly in the absence of national specifications, local authorities have employed their own methods of achieving the result they desire. One successful method has been by the use of an "approved list" derived from experience in road trials. An example was the approved list of polish-resistant aggregates used by the Greater London Council for many years. Problems had been experienced in specifying by minimum PSV alone. This may have been due to the anomalous behaviour of some aggregate types or possibly to the failure of some of the supplied aggregates to meet the standard claimed. Aggregates known to provide consistently good performance on the road were included in the list and others were added if road trials proved satisfactory, conversely aggregates giving a disappointing performance were deleted from the list. This system has advantages of simplicity and of avoiding any problems resulting from delays in testing. Disadvantages included the exclusion of many suitable materials and the work involved in monitoring where the aggregates were used and their performance.

A more complex system applicable to all road contracts has also been used. Here the orders for a year are scaled to the extent to which the quality was met during the previous year. A substandard product meant a smaller order in the following year and a producer of good materials was rewarded by a larger order.

Nowadays the general trend is to make as much use as possible of national specifications and only depart from them under special circumstances. This procedure has been eased by modification of the specification to allow additional aggregates to be used if known to be satisfactory. A rider is often included in the specification to the effect that other aggregates may be used with the approval of the engineer; sometimes such a rider is included in the British Standard specification.

8.6.1.4 The special problem of high-PSV aggregates

The most demanding requirements for aggregates are those to be used in skid-resistant road surfacings. The premium natural aggregates have a composition that is a fine balance between high resistance to polishing and adequate abrasion resistance (and strength). There is a general lowering of abrasion resistance as the PSV of gritstones (probably the only group that can provide the premium aggregates) increases and the best compromise is achieved with aggregates with a PSV of about 68 combined with an AAV of about 10. Those with higher PSVs are usually of inferior abrasion resistance and those with a lower AAV are usually more prone to polish. Although there is a linear relationship between the skid-resistance achieved and the PSV of the aggregate used (other factors such as traffic density and chipping size being equal) specifications for PSV have traditionally been framed in steps of 5. (An exception was the T2/67 minima of 59 and 62 respectively for different site categories, where the smaller step of 3 units was forced by a shortage of suitable aggregates at the time). This tradition probably arose during the early days of the roughness number test (roughness numbers were in steps of 5 units of "PSV"). Later when the concept of risk rating was proposed, steps of 5 in PSV were drawn up to correspond to the series of risk ratings. Although this step of 5 has been found to work reasonably satisfactorily, some problems of supply have been encountered at the upper end of the range, as with the T2/67 figures.

Experience has shown that the 65 to 69 range is the highest that is being regularly achieved with natural aggregates. In view of the scarcity of such gritstones it might prove prudent to employ smaller steps in specifying suitable aggregates for particular conditions. This would ensure that the very best aggregate would be "reserved" for use at the sites where it is most needed.

8.6.2 Skid-resistance maintenance policies

Maintenance policies that incorporate skid-resistance standards are desirable, but implementation presents serious problems. Of these the financial and legal implications are of most concern. Financial constraints mean that the best skid-resistant materials cannot be afforded for all roads, even if there was sufficient quantity. Consequently the funds that are available have to be proportioned to the needs of the different road sites. Legal constraints mean that the publication of minimum values could lead to claims for compensation on sites where the value for some reason or another falls below that published. At first sight this appears to be reasonable. However implementation of this type of standard involves a large amount of work in monitoring the road networks, all of which cannot be carried out at the same time. Additionally there is inherent variability of the skid-resistance characteristics of a road and factors such as a spell of very hot weather can cause a rapid lowering of skid-resistance. Thus, while it would be possible to maintain most of the roads in a prescribed condition all the time (assuming adequate funds to be available), and all of the roads for most of the time, it would be virtually impossible to maintain all the roads at the prescribed level all the time.

The influence of these and other factors have held back implementation of skid-resistance standards in all countries even though there have been a number of different approaches to the problem. These can be classed into three levels of action.

8.6.2.1 level 1

This is the simplest approach. Action is only taken when the accident level increases at a site. When such an increase is reported, measurements of skid-resistance are made (usually with the portable skid-resistance tester) and a decision is then made as to whether skid-resistance is playing a significant part in causing the increase in accidents. If the finding is positive the appropriate road authorities can give priority to resurfacing.

8.6.2.2 level 2

This level has been employed by several countries. It involves a general programme of road improvement in which the surface characteristics of the roads are progressively improved towards a uniform and generally acceptable level of skid-resistance appropriate to the different classes of road and road-site. The degree of success of these policies depends upon the funds made available to the road authorities and their proportionment among the many calls upon them.

A policy that has been found to be reasonably effective and, at the same time be subject to uniform interpretation in litigation, is to set mandatory requirements for the materials used in surfacing contracts. These will have been calculated to give the desired skid-resistance. Such a policy cannot be completely satisfactory as it does not take account of the deterioration, which may be gradual or sudden, in skid-resistance.

This compromise level of action has been successfully employed in Great Britain for many years. Non-mandatory target levels of skid-resistance have been proposed and mandatory requirements have been specified for new surfacings (examples are minimum polished-stone values of the aggregate and minimum macro-texture depths).

8.6.2.3 level 3

The third level involves both skid-resistance "standards" and new surfacing standards. A policy of this type has been implemented in the UK's Department of Transport. The policy documents have been published in Departmental Standard HD 15/87 (Department of Transport, 1987a) and the accompanying Advice Note HA 36/87 (Department of Transport, 1987b). Details of the skid-resistance levels of these standards are given in Chapter 7. In essence these documents detail procedures for making surveys with SCRIM on in-service trunk roads and outline appropriate remedial action where the investigatory levels dictate. These includes the erection of warning signs until the site has been investigated and appropriate action taken.

A similar policy has been included in the Local Authorities Associations "Code of Practice". While not mandatory this document gives recommendations to Local Authorities.

The more general adoption of these policies will yield significant benefits in accident reduction and it is hoped that they will also provide a means of assessing the priorities for limited funding.

8.6.3 Further savings in accidents and their cost by improvements in skid-resistance

The average cost of a personal injury accident in 1986 was £15,840 (See 6.3.2 above) and the number of accidents that would be saved by an increase in SCRIM coefficient of 0.01 was estimated to be 1,822 (See 6.3.3 above). It follows that the estimated national saving in accident cost for the year for each 0.01 increase in SCRIM coefficient would be nearly £29m.

The above estimate is based on average values of SCRIM coefficient (about 0.45), and therefore gives only an overall estimate of the savings. The work of Giles (Giles, 1957) showed that skidding accidents were infrequent where the coefficient was above 0.60, the risk increased sharply by more than 20 times as the coefficient fell to values of 0.40-0.45 and by about 300 times when the coefficient was 0.30-0.35. It follows that the improvement given by unit increase in SCRIM coefficient would be greater on low-coefficient roads and less on high-coefficient roads.

Similarly the severity of accidents tends to be greater on the higher speed roads leading to a higher proportion of serious injuries and fatalities. Hence the benefit per accident saved would be more on high speed roads and less in built up areas.

To these costs should be added the cost of non-injury accidents which do not require reporting to the police and for which there are no statistics. An estimate has been made of the cost of damage only accidents (see 6.3.2 above); this cost is only £658 as compared with £15,840 for the average injury accident. Nevertheless the saving that would follow an improvement in SCRIM coefficient would be considerable because of the large number of this type of accident.

The Road Research Laboratory's Laboratory Report 79 (Dawson, 1967) gives information on the cost and frequency of damage-only accidents. Information from 15 insurance companies showed that the ratio of damage only accidents to personal injury accidents was 1:7.4 for cars, 1:20.3 for goods vehicles, 1:3.4 for public service vehicles and 1:1.3 for motor cycles. The overall ratio was 1:6.8. Assuming that the benefit of improved SCRIM coefficient is proportionally the same as with injury accidents, the annual saving of damage only accidents per unit increase in SCRIM coefficient is estimated to be 1,822 x 6.8 x £658 = approximately £8m.

The above estimates are global figures derived by the Author and do not relate to specific types of road site. However a specific estimate has been made by the Department of Transport of the reduction in road casualties that can be achieved by improving the skid-resistance of our trunk road network (see Chapter 6.3.5). The striking way in which the Department of Transport's strategy has not only saved lives, but has also been very cost effective, gives a clear indication of the likely benefits of more widespread implementation of skidding standards.

8.6.4 Tougher standards for tougher chippings

When the PSV and other requirements for surfacing aggregates were first set by the Ministry of Transport, it was agreed that unnecessary testing should be avoided. Consequently when strength was being considered, it was decided not to set a requirement because experience had shown that all known aggregates of adequate abrasion resistance were also sufficiently strong.

The recent increased demand for high macro-texture has led to the need for the firmer embedment of chippings in hot-rolled asphalt and higher rates of spread. Stiffer asphalts are being used which result in greater stressing of the chippings during rolling and subsequent trafficking. The earlier decision should be reviewed: it might now be desirable to include a minimum strength requirement for chippings. A similar need has already arisen for chippings for surface dressing concrete, where rolling on the rigid substrate requires tougher chippings.

Acknowledgements

This review was prepared under contract to the Transport and Road Research Laboratory. The Author wishes to particularly thank the Project Officer, Mr. D M Colwill, Head of the Materials and Construction Division of the TRRL, for his encouragement and generous assistance. He also wishes to thank the many people both at the TRRL and the Highway Engineering Division of the Department of Transport who have given advice and help.

Grateful thanks are also due to my friend Dennis Ayres, a layman in Highway Engineering matters, whose comments have helped to make the text more readable. This also applies to my wife, Jean, whose patience and back-up have made the project possible.

References

ABRAMS, D A, 1919. *Engineering News Record*, 182.

ARUP, O, 1991. Occurrence and utilisation of mineral and construction waste. Department of the Environment.

ASTM - see Chapter 7 for ASTM standards.

AMERICAN SOCIETY FOR TESTING MATERIALS, 1901 - 1990. (See Chapter 7 for ASTM standards).

BACMI, 1987. Marine Dredged Aggregates. British Aggregate Construction Mateials Industries, London.

BATURIN, V P, 1942. A new method for studying sandy silt sediments, granulo-morphological analysis. *Comptes Rendus de L'Academie des Sciences de l'URSS*. 37(2), 66-68.

BECKMANN, D, 1964. Zur Bewertung der Griffigkeit von nassen Farhbahnoberflachen - II. Teil: Telative Unfallhaufigkeit bei Nasse. *Straβe und Autobahn*, 1964, 15 (6), 205/210.

BIRD, G and W J O SCOTT, 1936. Studies in road friction. I. Road surface resistance to skidding. *Department of Scientific and Industrial Research, Road Research Technical Paper No 1*. H M Stationery Office. London.

BIRD, G and R A Miller, 1937. Studies in road friction. II. An analysis of the factors affecting measurement. *Department of Scientific and Industrial Research, Road Research Technical Paper No 2*. H M Stationery Office. London.

BRADLEY, J and R F ALLEN, 1930-1. Factors affecting the behaviour of rubber-tyred wheels on road surfaces. *Proc. Inst. Auto. Engrs*. 1930-31, 25, 63-82.

BRENGARTH and LABORDE, 1977. Le Drainoroute. *Bulletin de Liaison de Laboratoires de Ponts et Chaussees No 87*.

BRITISH GEOLOGICAL SURVEY, 1985. Directory of Mines and Quarries. British Geological Survey, London.

BRITISH STANDARDS INSTITUTION, 1901 - 1990. (See Chapter 7 for BS standards).

BROWN, J R, 1967. An experiment comparing the performance of road stones used as chippings in rolled asphalt. *Ministry of Transport, RRL Report LR 63*. Road Research Laboratory. Crowthorne.

BROWN, J R, 1973. Pervious bitumen-macadam surfacings laid to reduce splash and spray at Stonebridge, Warwickshire. *Department of the Environment, TRRL Laboratory Report LR 563*. Transport and Road Research Laboratory. Crowthorne.

BROWN, J R, 1977. A full-scale experiment to compare surfacings for maintaining bituminous roads, A1 Buckden. *Department of Transport, TRRL Supplementary Report SR 323*. Transport and Road Research Laboratory, Crowthorne.

BROWN, J R, 1986. The performance of surfacings for maintaining bituminous roads: A1 Buckden. *Department of Transport, TRRL Research Report RR 70*. Transport and Road Research Laboratory, Crowthorne.

BS - see Chapter 7 for British Standards.

BUILDING RESEARCH STATION, 1963. Shrinkage of natural aggregates in concrete. *Building Research Station Digest No 35*. H M Stationery Office. London.

BULLAS, J C and G WEST, 1991. Specifying clean hard and durable for bitumen macadam roadbase. *Department of Transport, TRRL Research Report RR 284.* Transport and Road Research Laboratory, Crowthorne.

BURNS, J, 1978. The use of waste and low-grade materials in road construction: 6. Spent oil shale. *Department of the Environment, TRRL Laboratory Report LR 818.* Transport and Road Research Laboratory. Crowthorne.

CARMAN, P C, 1938. Determination of the specific surface of powders. *J.Soc.Chem.Ind.*, 57(7), 225-234.

CHRISTIE, A W, 1966. Research on street and highway lighting with particular reference to their effect on accidents. *The Institution of Municipal Engineers, Annual Conference, 1966.* The Institution of Municipal Engineers.

CLEMMER, H F, 1943. Report on significance of tests of concrete and concrete aggregates. *American Society for Testing Materials,* Philadelphia.

COOPER, D R C, 1974. Measurement of road surface texture by a contactless sensor. *Department of the Environment, TRRL Laboratory Report LR 639.* Transport and Road Research Laboratory. Crowthorne.

CRONEY, D and J C JACOBS, 1967. The frost susceptibility of soils and road materials. *Ministry of Transport, RRL Laboratory Note LR 90.* Road Research Laboratory. Crowthorne.

DAINES, M E, 1985. Pervious macadam trials on a trunk road A38 Burton bypass, 1984. *Department of Transport, Research Report RR 57.* Transport and Road Research Laboratory, Crowthorne.

DAWSON, R F F, 1967. Cost of Road Accidents in Great Britain. *Ministry of Transport, RRL Report LR 79.* Road Research Laboratory, Crowthorne.

DEPARTMENT OF TRANSPORT, 1976. Specification requirements for aggregate properties and texture depth for bituminous surfacings to new roads. *Technical Memorandum H 16/76.* Department of Transport.

DEPARTMENT OF TRANSPORT, 1986/8. Specification for Highway Works. H M Stationery Office, London.

DEPARTMENT OF TRANSPORT, 1987a. Skidding resistance of in-service trunk roads. *Departmental Standard HD 15/87.* Department of Transport.

DEPARTMENT OF TRANSPORT, 1987b. Skidding resistance of in-service trunk roads. *Advice Note HA 36/87.* Department of Transport.

DEPARTMENT OF TRANSPORT, 1987c. Road Accidents: Great Britain 1986: The Casualty Report. *Department of Transport, Scottish Development Department, Welsh Office.* Her Majesty's Stationery Office.

DEPARTMENT OF TRANSPORT, 1988. Improved skidding resistance on roads will save lives. *Department of Transport Press Notice No.24,* 9th January 1988. Department of Transport, London.

DIRECTOR OF ROAD RESEARCH, 1935-1966. Road Research 1935-1939; 1946-1966 (*Annual Reports*). Road Research Laboratory. H M Stationery Office. London.

DOE, 1989. Guidelines for aggregate provision in England and Wales. *Department of the Environment and Welsh Office MP*G 6. HM Stationery Office, London.

DUNAGAN, W M, 1940. The application of some of the newer concepts to the design of concrete mixes. *J.Amer.Conc.Inst.* 11(6), 649-684.

EDWARDS, L N, 1918. *Proc. ASTM,* 118, 235.

FRANKLIN, R E, 1978. The skidding resistance of concrete: development of a polishing test for fine aggregate. *Department of the Environment, TRRL Supplementary Report SR 444.* Transport and Road Research Laboratory. Crowthorne.

FRANKLIN, R E and A J J CALDER, 1974. The skidding resistance of concrete: the effect of materials under site conditions. *Department of the Environment, TRRL Laboratory Report LR 640*. Transport and Road Research Laboratory. Crowthorne.

GEE, S, W L KING AND R R HEGMAN, 1974. Pavement texture measurement by laser: a feasibility study. *ASTM Conf. Surface texture versus skidding STP 583*, 29-40.

GEOLOGICAL SOCIETY, 1985. Aggregates: sand, gravel and crushed rock aggregates for constructional purposes. *Geological Society Engineering Group, Special Publication No.1*. The Geological Society, London.

GILES, C G, 1957. The skidding resistance of roads and the requirements of modern traffic. *Proc. Inst. Civ. Eng*: 6, 216-249.

GILES, C G, B E SABEY and K H F CARDEW, 1964. Development and performance of the portable skid-resistance tester. *Department of Industrial and Scientific Research, RRL Technical Paper No 66*. H M Stationery office. London.

GILLOTT, J E, M A DUNCAN and E G SWENSON, 1973. Alkali-aggregate reaction in Nova Scotia. Part IV. Character of the reaction. *Chem., Concr. Res.* 3, 521-535.

GLOVER, M L, 1976. A network flow model of the distribution of aggregates under free competition. *Department of the Environment, TRRL Supplementary Report SR 208*. Transport and Road Research Laboratory. Crowthorne.

GLOVER, M H and B A SHANE, 1979. The distribution of aggregates - results of computer modelling studies. *Department of the Environment, TRRL Laboratory Report LR 898*. Transport and Road Research Laboratory. Crowthorne.

GRAF, O, 1936. Betonstraßenbau und Material prufung. *Straße*, 3,52-6.

GREEN, E H, 1974. Black deposits on motorways, Interim report to December 1973. *Department of the Environment, TRRL Laboratory Supplementary Report SR 74 UC*. Transport and Road Research Laboratory. Crowthorne.

GUTT, W and R J COLLINS, 1987. Sea dredged aggregates in concrete. *Department of the Environment, BRE Information Paper, IP 7/87*. Building Research Establishment, Garston.

HANSEN, W C, 1944. Studies relating to the mechanism by which the alkali-aggregate reaction produces expansion in concrete. *J. Amer. Conc. Inst.*, 15, (3), 213-27.

HATHERLEY, L W and D R LAMB, 1970. Accident prevention in London by road surface improvements. *Paper to the 6th World Highway Conference, Montreal*.

HAWKES, J R and J R HOSKING, 1972. British arenaceous rocks for skid-resistant road surfacings. *Department of the Environment, TRRL Laboratory Report LR 488*. Transport and Road Research Laboratory. Crowthorne.

HAWKINS, *et al*, 1983. Alkali-aggregate reaction - minimising the risk of alkali-silica reaction - Guidance Note. *Report of a Working Party*, 97-304. Cement and Concrete Association.

HAYES, G G and D L IVEY, 1974. Texture, skid resistance and the stability of automobiles in limit manoeuvres. *ASTM Conf. Surface textures versus skidding STP 583*, 127-139.

HENRY, J J and R R HENGMAN, 1974. Pavement texture measurement and evaluation. *ASTM Conf. Surface texture versus skidding. STP 583*, 3-17.

HEYSTRAETEN, G Van, 1975. La rugosite des revetements - Le beton de ciment cloute. *La Technique Routiere*. XX/4, 1-20.

HOPKINS, L C, 1959. The problems of providing aggregates for road surfacing materials which have good non-skid properties. *Inst. Highway Engrs*. Winchester.

HOSKING, J R, 1955. A comparison of the tensile strength, crushing strength and elastic properties of eleven roadmaking rocks. *The Quarry Manager's Journal*, 39, (4), 200-11.

HOSKING, J R, 1961. An investigation into some factors affecting the results of bulk density tests for aggregates. *Cement, Lime, Grav.*, 36(11), 319-26.

HOSKING, J R,, 1964. The effect of amplitude and frequency of vibration on the performance of a vibratory screen. *Quarry Managers' J.*, 48(9), 349-51.

HOSKING, J R,, 1965. Measurement of factors affecting screening efficiency. Methods for use under working conditions. *Quarry Managers' J.*, 49(4), 161-5.

HOSKING, J R, 1967a. Roadstone tests on present-day blastfurnace slag. *Ministry of Transport, RRL Report LR 96.* Road Research Laboratory. Crowthorne.

HOSKING, J R, 1967b. An experiment comparing the performance of road stones in different bituminous surfacings; A30 Blackbushe, Hants. *Ministry of Transport, RRL Report LR 81.* Road Research Laboratory. Crowthorne.

HOSKING, J R, 1968. Factors affecting the results of polished-stone value tests. *Ministry of Transport, RRL Report LR 216.* Road Research laboratory. Crowthorne.

HOSKING, J R, 1970. Synthetic aggregates of high resistance to polishing: Part 1 - gritty aggregates. *Ministry of Transport, RRL Report LR 350.* Road Research Laboratory. Crowthorne.

HOSKING, J R, 1973. The effect of aggregate on the skidding resistance of bituminous surfacings: factors other than resistance to polishing. *Department of the Environment, TRRL Laboratory Report LR 553.* Transport and Road Research Laboratory. Crowthorne.

HOSKING, J R, 1974. Synthetic aggregates of high resistance to polishing: Part 3 - porous aggregates. *Department of the Environment, TRRL Laboratory Report LR 655.* Transport and Road Research Laboratory. Crowthorne.

HOSKING, J R, 1976. Aggregates for skid-resistant roads. *Department of the Environment, TRRL Laboratory Report LR 693.* Transport and Road Research Laboratory, Crowthorne.

HOSKING, J R, 1986. Relationship between skidding resistance and accident frequency: estimates based on seasonal variation. *Department of Transport, TRRL Research Report RR 76.* Transport and Road Research Laboratory, Crowthorne.

HOSKING, J R and F A JACOBS, 1974. Synthetic aggregates of high resistance to polishing: Part 4 - specially shaped aggregates aggregates. *Department of the Environment, TRRL Laboratory Report LR 656.* Transport and Road Research Laboratory. Crowthorne.

HOSKING, J R and D C PIKE, 1985. The methylene blue dye adsorption test in relation to aggregate drying shrinkage. *J.Chem.Tech.Biotechnology.* 35A, 185-194.

HOSKING, J R and DIANE RITSON, 1968. The measurement of colour of roadstones. *Ministry of Transport, RRL Report LR 158.* Road Research Laboratory. Crowthorne.

HOSKING J R, P G ROE and L W TUBEY, 1987. Measurement of the macro-texture of roads. Part 2: A study of the TRRL mini texture meter. *Department of Transport, TRRL Research Report RR 120.* Transport and Road Research Laboratory. Crowthorne.

HOSKING J R and P B STILL, 1979. Measurement of the macro-texture of roads. Part 1: Trials with the contactless sensor on motorway M1. *Department of the Environment, Department of Transport, TRRL Supplementary Report SR 498.* Transport and Road Research Laboratory. Crowthorne.

HOSKING, J R and W SZAFRAN, 1968. The use of synthetic resins to speed-up the polished-stone value determination. *Roads and Road Construction,* 46(549),277-8.

HOSKING, J R and L W TUBEY, 1969. Research on low-grade and unsound aggregates. *Ministry of Transport, RRL Report LR 293.* Road Research Laboratory. Crowthorne.

HOSKING, J R and L W TUBEY, 1972. Aggregates for resin-bound skid-resistant road surfacings. *Department of the Environment, TRRL Laboratory Report LR 466.* Transport and Road Research Laboratory. Crowthorne.

HOSKING, J R and L W TUBEY, 1973. Experimental production of calcined bauxites for use as road aggregates. *Department of the Environment, TRRL Laboratory Report LR 588.* Transport and Road Research Laboratory. Crowthorne.

HOSKING, J R and L W TUBEY, 1974. Effect of turning and braking on the polishing of roadstone by traffic. *Department of the Environment, TRRL Supplementary Report SR 103 UC.* Transport and Road Research Laboratory. Crowthorne.

HOSKING, J R and G C WOODFORD, 1976a. Measurement of skidding resistance. Part I. Guide to the use of SCRIM. *Department of the Environment, TRRL Laboratory Report LR 737.* Transport and Road Research Laboratory, Crowthorne.

HOSKING, J R and G C WOODFORD, 1976b. Measurement of skidding resistance. Part II. Factors affecting the slipperiness of a road surface. *Department of the Environment, TRRL Laboratory Report LR 738.* Transport and Road Research Laboratory, Crowthorne.

HOSKING, J R and G C WOODFORD, 1976c. Measurement of skidding resistance. Part III. Factors affecting SCRIM measurements. *Department of the Environment, TRRL Laboratory Report LR 739.* Transport and Road Research Laboratory, Crowthorne.

HOSKING, J R and G C WOODFORD, 1978. Measurement of skidding resistance. Part IV. The effect on recorded SFC of design changes in the measuring equipment. *Department of the Environment, TRRL Supplementary Report SR 346.* Transport and Road Research Laboratory, Crowthorne.

HUGHES, B P and B BAHRAMIAN, 1966. A laboratory test for determining the angularity of aggregate. *Mag.Conc.Res.*, 18(56), 147-152.

HUGHES, B P and B BAHRAMIAN, 1967. An accurate laboratory test for determining the absorption of aggregates. *Materials Research and Standards.*, 7(1), 18-23.

ICHIHARA, K, 1973. Communication "Relationship between skidding resistance and accidents". 1973.

INTERNATIONAL STANDARDS ORGANISATION. ISO: 5725. Precision of tests methods - Determination of repeatability and reproducibility by inter-laboratory tests.

JACOBS, F A, 1983. M40 High Wycombe By-Pass: Results of a bituminous surface texture experiment. *Department of the Environment, Department of Transport, TRRL Laboratory Report LR 1065.* Transport and Road Research Laboratory. Crowthorne.

JAMES, J G, 1960. A small-scale road experiment to compare various aggregates for use in resin-based surface dressings. *RRL Research Note No RN 3843.* Road Research Laboratory. Harmondsworth (Unpublished).

JAMES, J G, 1963. Epoxy resins as binders for road and bridge surfacings. *Roads and Road Construction.* 41, (488). 23643.

JAMES, J G, 1965. Light-coloured rolled asphalt. *Roads and Road Construction*, 43(514), 309-312.

JAMES, J G, 1968. Calcined bauxite and other artificial polish-resistant roadstones. *Ministry of Transport, RRL Report LR 84.* Road Research Laboratory. Crowthorne.

JAMES, J G, 1971. Trial of Epoxy-Resin/Calcined-Bauxite surface dressing on A1, Sandy, Bedfordshire, 1968. *Department of the Environment, TRRL Laboratory Report LR 381.* Crowthorne, 1971. Road Research Laboratory.

JAMES, J G, 1972. Quantities and prices in road construction, 1969. A brief analysis of 60 successful tenders. *Department of the Environment, TRRL Laboratory Report LR 513.* Transport and Road Research Laboratory. Crowthorne.

JAMES, J G and J R HOSKING, 1967. Improvements in or relating to artificial roadstone. *British Patent No.38630/67.*

JAMES, J G and D R LAMB, 1974. Developments in resin-based skid-resistant road surfacings. *International Road Federation Regional Conference, Budapest*, 1974.

KAPLAN, M F, 1958. The effect of the properties of coarse aggregate on the workability of concrete. *Mag., Conc. Res.* 10(29), 63-74.

KNAPP, R, circa 1975. A study of the correlation between vehicle speed and length of skid mark. *Aylesbury College of Further Education. Aylesbury.* (Unpublished).

KNIGHT, B H, 1935. Road aggregates, their use and testing. Edward Arnold & Co., London.

KNILL, D C, 1960. Petrographic aspects of the polishing of natural roadstones. *J.Appl.Chem.*,10(1),28-35.

LEE, A R, 1974. Blastfurnace and steel slag: Production, properties and uses. Edward Arnold. London.

LEE, A R and J H NICHOLAS, 1954. Adhesion and adhesives, fundamentals and practice. *Soc.,Chem.,Ind.* London.

LEES, G, 1961. A new method for determining the angularity of particles and some general observations on the analysis of their shape. *Proc.,Midland Soil Mechanics and Foundation Eng.Soc.*, 4, Paper No. A7.

LEES, G, 1964. The measurement of particle elongation and flakiness: a critical discussion of British Standard and other methods. *Mag.Conc.Res.*, 16(49), 225-230.

LEES, G and I D KATEKHDA, 1974. Prediction of medium and high speed skid resistance values by means of a newly developed outflow meter. *Proc.Assoc. Asphalt Paving Technologists,* 43, 436-464.

LEYDER, J P, 1965. La rugosite des revetements en beton de ciment.*La Technique Routiere.* X/3, 3-26.

LOCAL AUTHORITY ASSOCIATIONS, 1989. Highway Maintenance - A code of good practice. Association of County Councils, Association of District Councils, Association of Metropolitan Authorities and Convention of Scottish Local Authorities. Association of County Councils (on behalf of Local Authority Associations).

LOVEGROVE, E J, J A HOWE, and J S FLETT, 1929. Attrition tests of British roadstones. *Mem. Geol. Surv. UK.* H M Stationery Office. London.

LUPTON, G N, 1968. Field testing of skidding. *Proc. Symp. "The influence of the road surface on skidding".* University of Salford.

LUPTON, G N and T WILLIAMS, 1973. Study of the skid resistance of different tire tread polymers on wet pavements with a range of surface textures. *ASTM Special Technical Publication.* American Society for Testing Materials,

MACLEAN, D J and F A SHERGOLD, 1958. The polishing of roadstone in relation to the resistance to skidding of bituminous road surfacings. *Department of Scientific and Industrial Research, Road Research Technical Paper No.43.* H M Stationery Office. London.

MAIR, I J, 1934. *Inst. C E*, Selected eng. papers.

MAJCHERCZYK, R 1974. Influence de la rugosite geometrique d'un revetement sur l'evacuation de l'eau a l'interface pneu-route et sur le derapage des vehicules. *Annales ITBTP No 318.*

MAJCHERCZYK, R and MARCHI, 1978. Experimentation sur les revetements permeables et drainants. *Annales ITBTP No 541.*

MARKWICK, A H D, 1936. The shape of road aggregate and its testing. *Department of Scientific and Industrial Research, RRL Road Research Bulletin No.2.* H M Stationery Office. London.

MARKWICK, A H D AND F A SHERGOLD, 1945. The aggregate crushing test. *J. Inst. C. Engrs..* 24, (6), 125-33.

MATHEWS, D H, 1958. Adhesion in bituminous materials: a survey of our present knowledge. *J., Inst., Petroleum*. 44(420), 423-432.

MATHEWS, D H and D M COLWILL, 1962. The immersion wheel tracking test. *J., Appl., Chem.*, Vol 12, 505-509.

MILLER, M M and H D JOHNSON, 1973. Effect of resistance to skidding on accidents: surface dressing on elevated section of M4 Motorway. *Department of the Environment, TRRL Laboratory Report LR 542*. Transport and Road Research Laboratory, Crowthorne.

MINISTRY OF TRANSPORT, 1963. Specification for road and bridge works. H M Stationery Office. London. (3rd Edition).

MINISTRY OF TRANSPORT, 1967. Polished stone values of aggregate for bituminous wearing courses and surface dressings. *Ministry of Transport Technical Memorandum No T2/67*. Ministry of Transport. London.

MINISTRY OF TRANSPORT, 1968. The use of colliery shale as filling material in embankments. *Technical Memorandum T4/68*. London. (Unpublished).

MINISTRY OF TRANSPORT, 1970. Report of the Committee of Highway Maintenance. H M Stationery Office. London.

MINOR, CARL E, 1959. Degradation of mineral aggregates. *ASTM Special Technical Publication No 277*.

MONCRIEFF, D S, 1953. The effect of grading and shape on the bulk density of concrete aggregates. *Mag.Conc.Res.*, 5(14), 67-70.

NATIONAL PHYSICAL LABORATORY, 1914. *Annual Report for 1913/1914*. H M Stationery Office. London.

NEVILLE, GWEN, 1972. Replica technique for the study of aggregates by scanning electron microscopy. *Department of the Environment, TRRL Laboratory Report LR 445*. Transport and Road Research Laboratory. Crowthorne.

NEVILLE, GWEN, 1974. A study of the mechanism of polishing of roadstones by traffic. *Department of the Environment, TRRL Laboratory Report LR 621*. Transport and Road Research Laboratory. Crowthorne.

NUNNY, R S and P C H CHILLINGWORTH, 1986. Marine dredging for sand and gravel. *Department of the Environment Minerals Division*. HM Stationery Office, London.

PELLOLI, R, 1977. Road surface characteristics and hydroplaning. *Proc. Conf. Skidding Accidents, Colombus, Ohio*. Transportation Research Record 624, 27-32.

PIARC. See PERMANENT INTERNATIONAL ASSOCIATION OF ROAD CONGRESSES.

PERMANENT INTERNATIONAL ASSOCIATION OF ROAD CONGRESSES, 1975. Technical Committee on Slipperiness and Evenness. *XV World Road Congress, Mexico*. Permanent International Association of Road Congresses. Paris.

PERMANENT INTERNATIONAL ASSOCIATION OF ROAD CONGRESSES, 1979. Technical Committee Report on Testing of Road Materials. *XVI World Road Congress, Vienna*. Permanent International Association of Road Congresses. Paris.

PERMANENT INTERNATIONAL ASSOCIATION OF ROAD CONGRESSES, 1983. Technical Committee Report on Testing of Road Materials. *XVII World Road Congress, Sydney*. Permanent International Association of Road Congresses. Paris.

PHEMISTER, J, E M GUPPY, A H D MARKWICK and F A SHERGOLD, 1946. Roadstone: geological aspects and physical tests. *Department of Scientific and Industrial Research, Road Research Special Report No. 3*. H M Stationery Office. London.

PICKEL, W and G ROTHFUCHS, 1938. Bewertung der Kornform von Edelsplitt. *Straßenbau*. 29(171), 273-277.

PIKE, D C, 1971. The drying properties of flint gravels. *Department of the Environment, RRL Laboratory Report LR 386.* Road Research Laboratory. Crowthorne.

PIKE, D C, 1972. Compactability of graded aggregates, 1. Standard laboratory tests. *Department of the Environment, TRRL Laboratory Report LR 447.* Transport and Road Research Laboratory. Crowthorne.

PIKE, D C, 1973. Shear-box tests on graded aggregates. *Department of the Environment, TRRL Laboratory Report LR 588.* Transport and Road Research Laboratory. Crowthorne.

PIKE, D C, 1979. Variability in grading results caused by standard sample-reduction techniques. *Department of Transport, TRRL Supplementary Report SR 489.* Transport and Road Research Laboratory. Crowthorne.

PIKE, D C, *et al*, 1990. Standards for aggregates. Ellis Harwood. Chichester.

PIKE, D C and S M ACOTT, 1975. A vibrating hammer test for compactability of aggregates. *Department of the Environment, TRRL Supplementary Report SR 140UC.* Transport and Road Research Laboratory. Crowthorne.

PIKE, D C, S M ACOTT and R M LEECH, 1977. Sub-base stability: a shear-box test compared with other prediction methods. *Department of the Environment, TRRL Laboratory Report LR 785.* Transport and Road Research Laboratory. Crowthorne.

POOLE, A B and P SATIRIPOULOS, 1981. Reactions between dolomitic aggregate and alkali pore fluids in concrete. *Q.,J.,Eng.,Geol.,London.* 13(4). 281-287.

ROAD RESEARCH LABORATORY, 1949. Wartime activities of the Road Research Laboratory. *Department of Industrial Research, RRL.* H M Stationery Office. London.

ROAD RESEARCH LABORATORY, 1965. Recommendations for road surface dressing. *Road Note No 39.* H M Stationery Office. London.

ROAD RESEARCH LABORATORY, 1966. Sources of white and coloured aggregates in Great Britain. *Road Note No 25.* H M Stationery Office. London.

ROAD RESEARCH LABORATORY, 1970. A guide to the structural design of pavements for new roads. *Road Note No 29.* H M Stationery Office. London.

ROAD RESEARCH LABORATORY, 1967-1971. Road Research 1967-1971 (*Annual Reports*). Road Research Laboratory. H M Stationery Office. London.

ROAD RESEARCH LABORATORY AND GEOLOGICAL SURVEY AND MUSEUM, 1948. Sources of road aggregate in Great Britain: a list of larger roadstone quarries and gravel pits. H M Stationery Office. London.

ROAD RESEARCH LABORATORY AND GEOLOGICAL SURVEY AND MUSEUM, 1950. Sources of road aggregate in Great Britain. H M Stationery Office. London.

ROAD RESEARCH LABORATORY and INSTITUTE OF GEOLOGICAL SCIENCES, 1968. Sources of Road Aggregate in Great Britain. H M Stationery Office. London (4th Edition).

ROAD RESEARCH LABORATORY AND MILITARY ENGINEERING EXPERIMENTAL ESTABLISHMENT, 1965. Methods for testing the performance of rock granulators. *Road Note No. 37.* H M Stationery Office. London.

ROE, P G, 1976. The use of waste and low-grade materials in road construction: 4. Incinerated refuse. *Department of the Environment, TRRL Laboratory Report LR 728.* Transport and Road Research Laboratory. Crowthorne.

ROE, P G, L W TUBEY and G WEST, 1988. Surface texture depth measurements on some British roads. *Department of Transport, TRRL Research Report RR 143.* Transport and Road Research Laboratory. Crowthorne.

ROE, P G and D C WEBSTER, 1984. Specification for the TRRL frost heave test. *Department of Transport, TRRL Supplementary Report SR 829.* Transport and Road Research Laboratory, Crowthorne.

ROGERS, M P and T GARGETT, 1991. A skidding resistance standard for the national road network. *Highways and Transportation.* Vol 38, No. 4., pp10-16.

ROSSLEIN, D, 1941. Steinbrecheruntersuchungen unter besonderer Berucksichtigung der Kornform. *Forschungsarbeiten aus dem Straβenwesen. 32.*

RYELL, J, J J HAJEK and G R MUSGROVE, 1976. Concrete pavement surface textures in Ontario - Development testing and performances. *Annual Meeting of the TRB.* Washington, DC.

SABEY, BARBARA E, 1958. Pressure distributions beneath spherical and conical shapes pressed into a rubber plane, and their bearing on coefficients of friction under wet conditions. *Proc.Physical Society,* 61, 979-988.

SABEY, BARBARA E, 1966. Road surface texture and the change in skidding resistance with speed. *Ministry of Transport, RRL Report LR 20.* Road Research Laboratory. Harmondsworth.

SABEY, BARBARA E and G N LUPTON, 1964. Friction on wet surfaces of tire-tread type vulcanates. *Rubber Chemistry and Technology,* 37(4), 878-893.

SABEY, BARBARA E, T WILLIAMS and G N LUPTON, 1970. Factors affecting the friction of tires on wet roads. *International Safety Conference Compendium.* Society of Automotive Engineers. New York.

SABINE, P A, J A MOREY AND F A SHERGOLD, 1954. The correlation of the mechanical properties and petrography of a series of quartz-dolerite roadstones. *J.appl.Chem.,*4(3),131-7.

SALT, G F, 1979. Skid-resistant road surfacings and tyre noise. *Proc. Inst. Civ. Engrs,* Part 1, 1979. 66 Feb., 115-125.

SALT, G F and W S SZATKOWSKI, 1973. A guide to levels of skidding resistance for roads. *Department of the Environment, TRRL Laboratory Report LR 510.* Transport and Road Research Laboratory. Crowthorne.

SCHAFFER, R J and P HIRST, 1930. The preparation of thin sections of friable and weathered materials by impregnation with synthetic resin. *Proc.Geol.Assn.,*41,32-43.

SCHIEL, F, 1941. Die Kornform der Betonzuschlagstoffe und ihre Profung. *Betonstraβe.* 16(12), 181-186.

SCHIEL, F, 1948. Determination of particle shape. Abhandlungen uber Bodenmechanik und Grundbau Forschungsgesellschaft fur das Straβenwesen. Bielefeld, (Erich Schmidt Verlag), 63-5.

SCHULZE, K H, J DAMES and H LANGE, 1974. Untersuchungen uber die Verkehrssicherheit bei Nasse. *Straβenbau und Straβenverkehrstechnik,* edited by Bundesminister fur Verkehr. Abt. Straβenbau. Bonn.

SHELBURNE, T E, 1940. Crushing resistance of surface treatment aggregates. *Purdue University Engineering Experiment Station, Research Series No. 76.* Purdue University. Lafayette, Indiana.

SHERGOLD, F A , 1946. The effect of sieve loading on the results of sieve analysis of natural sands. *J.,Soc.,Chem.,Ind.,* 65,245.

SHERGOLD, F A, 1948. A review of available information on the significance of roadstone tests. *Department of Scientific and Industrial Research, Road Research Technical Paper No.10.* H M Stationery Office. London.

SHERGOLD, F A, 1953a. The test for the apparent specific gravity and absorption of coarse aggregate. *J.Appl.Chem.,*3(3),110-7.

SHERGOLD, F A, 1953b. The percentage voids in compacted gravel as a measure of its angularity. *Mag.Concr.Res.,* 5(13), 3-10.

SHERGOLD, F A, 1953c. The effect of high temperatures on the strength of roadmaking aggregates. *Rds & Rd Constr.*, 31(366),161-3.

SHERGOLD, F A, 1954a. The effect of freezing and thawing on the properties of roadmaking aggregates. *Rds & Rd Constr.*, 32(381),274-6.

SHERGOLD, F A, 1954b. A study of single-sized gravel aggregates for roadmaking. *Department of Scientific and Industrial Research, Road Research Technical Paper No.30.* H M Stationery Office. London.

SHERGOLD, F A, 1959. A study of granulators used in the production of road making aggregates. *Department of Scientific and Industrial Research, Road Research Technical Paper No.44.* H M Stationery Office. London.

SHERGOLD, F A, 1963. A study of the variability of roadstones in relation to sampling procedures. *Quarry Managers' J.*, 47(1), 3-8.

SHERGOLD, F A and J R HOSKING, 1959. A new method of evaluating the strength of roadstone with particular reference to the weaker types used in road bases. *Rds.& Rd.Constr.*, 37(438),164-7.

SHERGOLD, F A and J R HOSKING, 1963. Investigation of test procedures for argillaceous and gritty rocks in relation to breakdown under traffic. *Roads and Road Construction*: 41, (492), 376-8.

SHERGOLD, F A and J R MANNING, 1953. The assessment of shape in roadmaking aggregates. *Rds & Rd Constr.*, 31(361), 4-7.

SHERWOOD, P T, 1974. The use of waste and low-grade materials in road construction: 1. Guide to materials available. *Department of the Environment, TRRL Laboratory Report LR 647.* Transport and Road Research Laboratory. Crowthorne.

SHERWOOD, P T, 1975a. The use of waste and low-grade materials in road construction: 2. Colliery shale. *Department of the Environment, TRRL Laboratory Report 649.* Transport and Road Research Laboratory. Crowthorne.

SHERWOOD, P T, 1975b. The use of waste and low-grade materials in road construction: 3. Pulverised fuel ash. *Department of the Environment, TRRL Laboratory Report LR 686.* Transport and Road Research Laboratory. Crowthorne.

SHERWOOD, P T, 1987. Wastes for imported fill. *ICE Construction Guides.* Thomas Telford. London.

SHERWOOD, P T AND D C PIKE, 1984. Errors in the sampling and testing of sub-base aggregates. *Department of the Environment, TRRL Supplementary Report SR 831.* Transport and Road Research Laboratory. Crowthorne.

SHERWOOD, P T, L W TUBEY and P G ROE, 1977. The use of waste and low-grade materials in road construction: 7. Miscellaneous wastes. *Department of the Environment, TRRL Laboratory Report LR 819.* Transport and Road Research Laboratory. Crowthorne.

SHUPE, J W and R W LOUNDSBURY, 1959. Polishing characteristics of mineral aggregates. *Purdue University, Civil Engineering Reprint No. 160.* Lafayette, Indiana.

SIBBICK, R G and G WEST, 1989. Examination of concrete from the M40 Motorway. *Department of Transport, TRRL Research Report RR 197.* Transport and Road Research Laboratory. Crowthorne.

STANTON, T E, O J PORTER, L C MEDER AND A NICOL, 1942. California experiences with the expansion of concrete through reaction between cement and aggregate. *J.Amer.Concr.Inst.*, 13, (3),209-36.

STATE ROAD LABORATORY, 1973. Memorandum on research in relation with slipperiness and evenness of roads in the Netherlands. February, 1973. Delft, Holland. (State Roads Laboratory, Holland).

STEWART, E T and L M McCULLOUGH, 1985. The use of the methylene blue test to indicate the soundness of road aggregates. *J. Chem. Tech. Biotechnol..* 35A, 161-167.

STILL, P B and M A WINNETT, 1975. Development of a contactless displacement transducer. *Department of the Environment, TRRL Laboratory Report LR 690*. Transport and Road Research Laboratory. Crowthorne.

SWEET, H S AND K B WOODS, 1942. A study of chert as a deleterious constituent in aggregates. *Purdue University, Engineering Experimental Station, Research Series No. 86*. Lafayette, Indiana.

SWEETMAN, N B, 1974. A pilot-scale study of slag/calcined bauxite aggregate of high polishing resistance. *Department of the Environment, TRRL Supplementary Report SR 105 UC*. Transport and Road Research Laboratory. Crowthorne.

SZATKOWSKI, W S and J R HOSKING, 1972. The effect of traffic and aggregate on the skidding resistance of bituminous surfacings. *Department of the Environment, TRRL Laboratory Report LR 504*. Transport and Road Research Laboratory. Crowthorne.

TABOR, D, C G GILES and BARBARA E SABEY, 1958. Friction between tyre and road. *Engineering*. 186 (4842) 832-42.

TRANSPORT AND ROAD RESEARCH LABORATORY, 1969. Instructions for using the portable skid-resistance tester. *Road Note No 27*. H M Stationery Office. London.

TRANSPORT AND ROAD RESEARCH LABORATORY, 1972-1979. Transport and Road Research 1972-1979 (*Annual Reports*). Transport and Road Research Laboratory. H M Stationery Office. London.

TRANSPORT AND ROAD RESEARCH LABORATORY, 1982. Terminology for use with SCRIM data analysis and reporting. *Department of the Environment, Department of Transport, TRRL Leaflet LF 915*. Transport and Road Research Laboratory. Crowthorne.

TRANSPORT and ROAD RESEARCH LABORATORY, 1990. Skidding-resistance standard cuts accident. *Department of Transport, TRRL Press Notice 32/90*. Transport and Road Research Laboratory, Crowthorne.

TUBEY, L W, 1978. The use of waste and low-grade materials in road construction: 5. China clay sand. *Department of the Environment, TRRL Laboratory Report LR 817*. Transport and Road Research Laboratory. Crowthorne.

TUBEY, L W and J R HOSKING, 1972. Synthetic aggregates of high resistance to polishing: Part 2 - Corundum-rich aggregates. *Department of the Environment, TRRL Laboratory Report LR 467*. Transport and Road Research Laboratory. Crowthorne.

TUBEY, W L and P G JORDAN, 1973. The reproducibility and repeatability of the PSV determination. *Department of the Environment, TRRL Laboratory Report LR 552*. Transport and Road Research Laboratory. Crowthorne.

VALLIS, J W M, 1978. Friction Courses in Warwickshire. *Shell Bitumen Review*, 57. June, 1978.

VERES, R E, J J HENRY and J M LAWTHER, 1974. The use of tyre noise as a measure of pavement macro texture. *ASTM Conf. Surface texture versus skidding STP 583*, 18-26.

VERNEY, R, 1976. Aggregates: the way ahead. Advisory Committee on aggregates, Department of the Environment, Scottish Development Department, Welsh Office.

WALSH, H N, 1933. *Proc. Inst. C E of Ireland*. 59. 275.

WEBSTER, D C and G WEST, 1989. The effect of additives on the frost-heave of a sub-base gravel. *Department of Transport, TRRL Research Report RR 213*. Transport and Road Research Laboratory, Crowthorne.

WELLER, D E, 1970. A review of the low-speed skidding resistance of a number of concrete roads containing various aggregates. *Department of the Environment, RRL Laboratory Report LR 335*. Road Research Laboratory. Crowthorne.

WELLER, D E and D P MAYNARD, 1970a. The use of an accelerated wear machine to examine the skidding resistance of concrete surfaces. *Department of the Environment, RRL Laboratory Report LR 333.* Road Research Laboratory. Crowthorne.

WELLER, D E and D P MAYNARD, 1970b. The influence of materials and mix design on the skid resistance value and texture depth of concrete. *Department of the Environment, RRL Laboratory Report LR 334.* Road Research Laboratory. Crowthorne.

WEST, G and R G SIBBICK, 1988a. Alkali-silica reaction in roads. *Highways,* 56,1936,19-24.

WEST, G and R G SIBBICK, 1988b. Petrographical comparison of old and new control stones for the accelerated polishing test. *Quarterly Journal of Engineering Geology,* 21, 375-378. London.

WEST, G and R SIBBICK, 1989a. Alkali-silica reaction in roads. *Highways,* 57,(1949), 9-14.

WEST, G and R SIBBICK, 1989b. A mechanism for alkali-silica reaction in concrete roads. *International conference on alkali-aggregate reaction, Kyoto, Japan.* 17-20, July 1989.

WILSON, D S, 1966. An experiment comparing the performance of road stones in surface dressings, A40, West Wycombe, Bucks. *Ministry of Transport, RRL Report LR 46.* Road Research Laboratory. Crowthorne.

WRIGHT, P J F, 1955. A method of measuring the surface texture of aggregate. *Mag.Concr.Res.,*7,(21),151-60.

WRIGHT, N, 1976. Surface dressing on concrete roads. *Department of the Environment, TRRL Laboratory Report LR 736.* Transport and Road Research Laboratory. Crowthorne.

YOUNG, A E, 1985. The potential for accident reduction by improving urban skid resistance levels. *PhD Thesis.* Queen Mary College, University of London.

ZELINA, J, 1973. Communication to PIARC "Relations entre la resistance au glissement et les accidents en Tchechoslovaquie". March, 1973.

ZINGG, T, 1935. Beitrag zur Schotteranalyse. *Schweizerische Mineralogische und Petrographische Mitteilungen.* 15,39-140.

Appendix: Abbreviations and Acronyms

AASHTO	American Association of State Highway and Transportation Officials
AAV	Aggregate Abrasion Value
ACMA	Asphalt and Coated Macadam Association
ACV	Aggregate Crushing Value
AFNOR	Association Francaise de Normalisation
AIV	Aggregate Impact Value
ASTM	American Society for Testing Materials
BACMI	British Aggregate Construction Materials Industries
BFC	Braking-Force Coefficient
BS	British Standard
BSI	British Standards Institution
C&CA	Cement and Concrete Association
CBER	Calcined-Bauxite/Epoxy-Resin (dressing)
CBR	California Bearing Ratio
CEN	Comite Europeen de Normalisation
CIE	Commission Internationale de l'Eclairage
DIN	Deutsche Institut fur Normung
DSIR	Department of Scientific and Industrial Research
DTp	Department of Transport
EEC	European Economic Community
EFTA	European Free Trade Association
ESC	Equilibrium SCRIM Coefficient
FBA	Furnace Bottom Ash
IGS	Institute of Geological Sciences
ISO	International Standards Organisation
JHPC	Japan Highways Public Corporation
MOT	Ministry Of Transport
MPVS	Maximum Proportion of Volume occupied by Solids
MSSC	Mean Summer SCRIM Coefficient
MVHT	Modified Vibrating Hammer Test
NBS	National Bureau of Standards
NCB	National Coal Board
NPL	National Physical Laboratory
PFA	Pulverised Fuel Ash
PIARC	Permanent International Association of Road Congresses
PMV	Polished-Mortar Value
PPV	Polished-Paver Value
PSC	Polished-Stone Coefficient
PSV	Polished-Stone Value
RILEM	Reunion Internationale des Laboratoires et de Recherches sur Materieux et les Constructions.
RRL	Road Research Laboratory

SAGA	Sand And Gravel Association
SC	SCRIM Coefficient
SCRIM	Sideway-force Coefficient Routine Investigation Machine
SFC	Sideway-Force Coefficient
SMTD	Sensor-Measured Texture Depth
SR	SCRIM Reading
SRT	Skid Resistance Tester
SRV	Skid Resistance Value
SV	Skid Value
TRRL	Transport and Road Research Laboratory
TWIT	Total Water Immersion Test

Index

AAV	24, 70, 86, 110, 119, 151	Bituminous mixture experiments	60
Abrasion, resistance to	46, 58, 70	Bituminous road construction	56
Abrasion value	86, 113	Bituminous surfacings	125
Accelerated wear machine	137	Black deposits	142
Accident rate	170	Blackbushe experiment	151
Accidents, saving in	174	Blastfurnace slag	15, 28, 51, 56, 130
Accident/skidding studies	171	Blastfurnace slag, chemical analysis of	101
Accuracy	72	Blastfurnace slag, dense	15
Acid rocks	58	Blastfurnace slag, light-weight	16
Acid-soluble material	100	Blastfurnace slag, lighter-weight	16
ACV	113	Blastfurnace slag, microscope test for	100
Adhesion to bituminous binders	20, 58	Blastfurnace slag, variability of	77
Aggregate abrasion test, (see also AAV)	70, 86	Braking-force coefficient, (see also BFC)	161
Aggregate crushing test	8, 83, 113, 117	Braking-force trailer	161
Aggregate crushing test, 1/4-standard	83	British Standard classification	104
Aggregate identification	70	British Standards, current	198
Aggregate impact test	84	British Standards Institution	5, 192
Aggregate impact test, modified	85	British Standards referring to aggregates	193
Aggregates for skid-resistance	125	Brittleness	85
Aggregates in concrete	137	BS 812:1938	194
Air-entrained concrete	48	BS 812:1943	195
AIV	110, 113	BS 812:1951	195
Alkali-aggregate reactivity	47	BS 812:1960	196
Alkali-carbonate reaction	47	BS 812:1967	196
Alkali-silica reaction	47-8	BS 812:1975	196
Alkali-silicate reaction	47	BS 812:1984-1990	197
Angularity	53, 91	Bulk density	51, 88
Angularity, degree of	90	Bulk sample	79
Angularity factor	91		
Angularity number	89, 115	CADAM classification	104
Appearance	59	Calcined bauxite	24-5, 65, 143, 151-3
Approved lists	210	Calcined-bauxite/epoxy-resin surfacings	142, 153
Arenaceous rocks	110	Calcined flint	59, 175
ASTM Standards	199	California bearing ratio	44, 103
Ballotini	59, 175	Cement-kiln dust	39
Basalts	19, 108	Cement stabilisation	36
Batch	79	Cementation test	101
Belgium	188	CEN Standards	192, 202
BFC	148, 153, 161-2, 172-3, 189-91	Chalk	40
Bitumen macadam	151	Chemical durability	47

Chemical tests	100	Dense tar surfacings	60, 151
China clay sand	38	Density	58, 87
Chipped concrete	146	Derby Ring Road experiment	149
Chippings for surface treatment	57	Deval attrition test	80-81, 113
Chippings, grading of	62	Deval attrition test, modified	81
Chlorides	18, 50, 100	Diabases	108
Classification	103, 204	Digest 35 shrinkage test	102, 122
Clay content	71	Diorites	108
Clay lumps	54, 68	Distribution of aggregates	32
Clay, silt and dust	53	Distribution of waste materials	32
Clean, hard and durable	59	Dolerites	49, 108
Cleanliness	58-9	Domestic refuse, incinerated	37
Coated macadam	56	Dorry abrasion test	86
Code of Good Practice	183	Drainoroute	167
Coefficient de polissage accelere	189	Dry attrition value	81, 113, 119
Coefficient of hardness	86	Dry-bound macadam	43
Coefficient of uniformity	116	Drying properties	19
Coefficient of wear	81	Drying shrinkage	20, 49, 102
Colliery shale	35	Dusting of slags	51, 100
Colliery shale, burnt	36		
Colliery shale, unburnt	35-6	Elevated section of M4 experiment	152
Colour	59, 103	Elongation coefficient	69
Committee Europeen de Normalisation	192	Elongation index	52, 92-3
Compactability	53-4, 94	Epidiorites	108
Compacting factor of concrete	54, 115	Equilibrium SCRIM coefficient	165
Compaction, pilot scale research	44	Exposed aggregate concrete	146
Compaction, research on	43		
Concrete, light-weight	17	Failure time	102
Concrete mix design	54	Falling of slags	51, 100
Concrete road construction	45	Fatigue strength tests	86
Concreting sand	18	Field settling test	95, 97
Cone granulators	29	Filler, bulk density of	89
Copper slag	17	Filler, void content of	71
Core-insert experiments	137, 148	Fine material	68
Correlation between tests	107	Fineness modulus	83
Crushed rock	56	Finland	188
Crushing	29	Flakiness coefficient	69
Crushing rolls	29	Flakiness index	52, 58, 69, 92-3
Crushing strength	82, 113, 117, 119	Flaky particles	203
Crushing under the roller	61-2, 82	Flexible road mixtures	56
Cumulative grading curve	95	Flexural strength of concrete	53
Czechoslovakia	172, 188	Flexural strength test	85
		Flint gravel	19
Decantation test	97	Flints	108
Decelerometer	161	Flow value of sands	70
Degradation factor	99	Fly ash	37
Demolition wastes	39	Foamed slag	16
Dense bituminous surfacings	60	Framework standards	192

France	189
Freezing and thawing	69, 98
Friable particles, test for	54
Friction course	146
Frost-heave	42, 99, 103
Frost susceptibility	37, 42, 57, 88, 99
Full-scale road experiments	148-9
Furnace clinker	40
Garrett Lane experiment	177
Germany, Federal Republic of	172
Giles proposals	178
Glass, waste	40
Grading	30, 57
Grading of aggregates	203
Grading of concrete aggregates	52
Granites	20, 107
Granular sub-base	42, 44
Granulators	29
Gravels	17, 56
Gritstones	22-3, 110, 152
Gypsum	49
Haddenham experiment	62, 149
Hassock	40
High-hysteresis rubber compounds	205
High-hysteresis rubber tyres	207
High-PSV aggregates	211
High-speed profilometer	167
High-speed road monitor	167
High-speed skid testing	157
High speed testing	161
High-speed texture meter	162, 168, 208
High temperatures	98
Hornfels	20
Hot-rolled asphalt	63
Igneous rocks	19, 112
Immersion wheel tracking test	102
Impact breakers	29
Impact value	84, 113
Incinerated domestic refuse	37
Indirect tensile strength test	86
International Standards Organisation	192
Investigatory levels of skid-resistance	181
Iron unsoundness	51, 100
Italy	172, 189

Japan	173, 190
Jaw granulators	29
Kingston By-Pass experiment	161
Knight, Dr. Bernard	6
Knight's classification	104
Laboratory sample	79
Light coloured aggregates	176
Lime unsoundness	51, 100
Limestone gravel	19
Limestones	21, 56, 108, 112
Limiting values	210
Local knowledge	210
Los Angeles abrasion test	68, 82, 113
Lovegrove attrition test	80, 113
Lovegrove, E J	4
Low-grade aggregates	34
Low-resilience tyres	154, 205, 208
Macadam, water-bound	41
Macro-texture	154, 157, 166, 181, 185, 205, 207-8
Macro-texture, measurement of	166
Magnesian limestones	21
Magnesium sulphate soundness test	57, 99
Marine deposits	18
Marshall Committee recommendations	178
Mean summer SCRIM coefficient	165
Metallurgical slags	14
Metamorphic rocks	20
Methylene blue dye adsorption test	71, 102, 122
Micro-Deval test	70, 81
Micro-texture	154
Mineralogical composition	70
Mini texture meter	168
Mining	201
Minority fraction	80
Mixtures of aggregates	131
Moisture content	52
Motor-cycle apparatus	161
National Skidding Resistance Survey	181
Netherlands, The	172, 190
Noise	174

Oil shale, spent	38
Optical methods.	203
Optimum binder content	88
Organic impurities tests	49
Outflow meters	208
Page impact test	84
Particle density	51, 68, 87
Particle shape	31, 52
Particle size distribution	67, 95
Particle-size tests, precision of	97
Pavers	65
Permanen International Association of Road Congresses	192
Pervious macadam	146
Petrography	107, 109
Petrological description	104
PFA	37
Phosphorus slag	17
Poland	190
Polish-resistance standards for aggregates	185
Polish-resistant aggregates	23
Polished-mortar value	93-4, 140
Polished-paver value	67, 94
Polished-stone coefficient	127
Polished-stone value, (see also PSV)	68, 93, 112, 119, 125-6, 128
Polishing resistance	58, 93, 109, 129, 147
Porosity	70
Porphyries	20, 108
PPVs	67
Precision	72, 74
Precision estimates, uses of	75
Production of aggregate	28
PSV	24, 110, 129-31, 137, 141, 143, 146-8, 151-2, 181, 185-6, 188-9, 191, 196
Pulverised fuel ash	36
Pyrites	49
Quarry wastes	40
Quartz-dolerites	108
Quartzites	20, 130
Ragstone	40
Railway ballast	81
Razor-blade shape	203
Reference datum	74

Reflecting properties of roads	175
Relative density	68
Repeatability	9, 72, 75
Reproducibility	9, 72, 75
Resin-bound surfacings	25, 65, 142-3
Resistance to skidding	154, 159, 178
Retardation	49
Rigden test	71
Risk rating	179
Road accidents, causes of	169
Road accidents, cost of	169
Road experiments	148
Road experiments, full-scale	10, 60
Road Research Laboratory, early work of	7
Road surface reflection characteristics	175
Roadbase	42
Roads, unsurfaced	41
Rolled asphalt	56, 149, 151
Rosslein gauge	92
Roughness factor	92
Roughness number	126-7
Round-hole sieves	67
Roundness	53
Roundness index	89
Rubber properties	205
Rubber, waste	40
Salt, de-icing	48
Sample increment	79
Sample reduction	69
Sampling	76
Sampling for sieve analysis	77
Sand equivalent test	68
Sand-patch method	166, 189
Sands	17
Sandstones	108
Sandy experiment	153
Scalping	29
Schists	20
Screening	31
Screening, efficiency of	31
Screening tests	74
SCRIM	131, 148, 156, 160, 162-3, 181, 207
SCRIM coefficient	165, 170, 183
SCRIM reading	165
Seasonal variation in skid-resistance	127, 163
Sedimentary rocks	21

Sedimentation test	97	Skiddometer BV5	191
Sensor-measured texture depth, (see also SMTD)	168	Slags, non-ferrous	39
		Slake durability test	99
Setts	65	Slate waste	39
SFC	126-8, 131, 146, 148, 151-2, 161-3, 170, 178-9, 185-6, 188	Slipperiness of the road surface	163
		Slurry seal	146, 153
SFC, index of	165	Small-scale road experiments	148-9
Shape	57, 69, 89	Smoothness	53
Shape coefficient	69	SMTD	168
Shaped aggregates	28	Sodium sulphate soundness test	57, 99
Shear-box test	43, 94	Soundness	57, 70, 98
Shear strength	43	Sources of road aggregate	14, 200
Shell content	50, 102	Spain	191
Shellgrip	143	Specific gravity	87
Sideway-force coefficient	178	Specifications for aggregates	192
Sideway-force Coefficient Routine Investigation Machine, (see also SCRIM)	162	Spent oil shale	38
		Spontaneous combustion	36
		Square-mesh sieves	67
Sideway force coefficient, (see also SFC)	160	SRV	141, 148, 161, 166, 183
		Staining	49
Sieve analysis	95	Steel slag	15, 17, 56
Sieve loading	95	Stewart impact test	84
Single-sized aggregates	18, 62	Stiffer asphalts	214
Skid mark, length of	208	Stopping distance	209
Skid-resistance	207	Stradographe	188-9
Skid-resistance, aggregates for	204	Streatham High Road experiment	149
Skid-resistance maintenance policies	211	Street lighting	175
Skid-resistance, overseas specifications for	188	Strength	46, 80
		Stripping	18-20, 71
Skid-resistance standard, Department of Transport's	181	Studded tyres	47, 71, 189
		Sub-base	42
Skid-resistance standard, Giles' proposals	178	Sulphate, total	101
		Sulphate, water-soluble	101
Skid-resistance standard, Local Authorities Code of Practice	183	Sulphates	49, 101
		Sulphur	51
Skid-resistance standard, Marshall recommendations	178	Sulphur unsoundness	51
		Super-quarries	201
Skid-resistance standard, TRRL's proposals	178	Surface characteristics	154, 205
		Surface dressing	63, 151
Skid resistance standards	178	Surface dressing on concrete	64
Skid-resistance surveys	181	Surface projection shape	157
Skid-resistance tester	161, 166, 183, 190-91	Surface roughness	91
Skid-resistance value, (see also SRV)	161, 166, 190	Surface texture	181
		Surface texture and noise	174
Skidding	131	Sweden	191
Skidding on dry roads	208	Switzerland	191
Skidding rate	170	Synthetic aggregates	25, 36, 204

Technical Memorandum H16/76	186
Ten-per-cent fines durability factor	57, 100
Ten-per-cent fines test	84
Ten-per-cent fines test, modified	84
Tensile strength	85, 113
Terminology for use with SCRIM data	163
Test portion	79
Test procedures	73
Tests for roadstones and aggregates	73
Texas ball mill test	99
Texture	89
Texture meters	167, 207
Thermal expansion of concrete	21
Tower Bridge	142
Tracked vehicles	46
Trade names	103
Traffic density	131
Traffic noise	174
Tread pattern	205
Turning and braking	137
Tyre hysteresis	205
Tyre pattern	205
Unsound aggregates	99, 102
Unsurfaced roads	41
Variability of aggregates	8, 76
Variability of grading results	79
Variation in skid-resistance at a road site	137
Vibrating hammer compaction test	43, 54, 94, 115, 117
Visibility and glare	175
Washington degradation test	99
Waste materials	34
Water absorption	51, 87
Water-bound macadam	41
Weathering	57, 70
West Wycombe experiment	151
Wet attrition value	81, 113
Wet-mix macadam	43
White aggregates	59, 175
Wind-screen damage	63
Workability of concrete	54
Zinc slag	17
Zingg's ratio	92